T0275702

How to Optimize Fluid Bed Processing Technology

How to Optimize Fluid Bed Processing Technology

Part of the *Expertise in Pharmaceutical Process Technology* Series

Dilip M. Parikh
President, DPharma Group Inc., Ellicott City, MD, United States

ACADEMIC PRESS
An imprint of Elsevier

Academic Press is an imprint of Elsevier
125 London Wall, London EC2Y 5AS, United Kingdom
525 B Street, Suite 1800, San Diego, CA 92101-4495, United States
50 Hampshire Street, 5th Floor, Cambridge, MA 02139, United States
The Boulevard, Langford Lane, Kidlington, Oxford OX5 1GB, United Kingdom

Notices
Knowledge and best practice in this field are constantly changing. As new research and experience broaden our
understanding, changes in research methods, professional practices, or medical treatment may become
necessary.

Practitioners and researchers must always rely on their own experience and knowledge in evaluating and using
any information, methods, compounds, or experiments described herein. In using such information or methods
they should be mindful of their own safety and the safety of others, including parties for whom they have a
professional responsibility.

To the fullest extent of the law, neither the Publisher nor the authors, contributors, or editors, assume any
liability for any injury and/or damage to persons or property as a matter of products liability, negligence or
otherwise, or from any use or operation of any methods, products, instructions, or ideas contained in the
material herein.

British Library Cataloguing-in-Publication Data
A catalogue record for this book is available from the British Library

Library of Congress Cataloging-in-Publication Data
A catalog record for this book is available from the Library of Congress

ISBN: 978-0-12-804727-9

For Information on all Academic Press publications
visit our website at https://www.elsevier.com/books-and-journals

Working together
to grow libraries in
developing countries

www.elsevier.com • www.bookaid.org

Publisher: Mica Haley
Acquisition Editor: Erin Hill-Parks
Editorial Project Manager: Molly McLaughlin
Production Project Manager: Punithavathy Govindaradjane
Designer: Mark Rogers

Typeset by MPS Limited, Chennai, India

DEDICATION

This book is dedicated to my dear friend **Mr. David M. Jones** who passed away in January 2016. David's contribution to the pharmaceutical industry and particularly in fluid bed process technology will be sorely missed.

Dilip M. Parikh
Ellicott City, MD, USA

CONTENTS

About the *Expertise in Pharmaceutical Process Technology* Series

Numerous books and articles have been published on the subject of pharmaceutical process technology. While most of them cover the subject matter in depth and include detailed descriptions of the processes and associated theories and practices of operations, there seems to be a significant lack of practical guides and "how to" publications.

The *Expertise in Pharmaceutical Process Technology* series is designed to fill this void. It will comprise volumes on specific subjects with case studies and practical advice on how to overcome challenges that the practitioners in various fields of pharmaceutical technology are facing.

FORMAT

- The series volumes will be published under the Elsevier Academic Press imprint in both the printed and electronic versions, with each volume containing approximately 100–200 pages. Electronic versions will be full color, while print books will be in black and white.

SUBJECT MATTER

- The series will be a collection of hands-on practical guides for practitioners with numerous case studies and step-by-step instructions for proper procedures and problem solving. Each topic should start with a brief overview of the subject matter, exposé and practical solutions of the most common problems with some case studies, and a lot of common sense (proven scientific rather than empirical practices).
- The series will try to avoid theoretical aspects of the subject matter and limit scientific/mathematical exposé (e.g., modeling, finite elements computations, academic studies, review of publications, theoretical aspects of process physics or chemistry) unless absolutely

vital for understanding or justification of practical approach as advocated by the volume author. At best, it will combine both the practical ("how to") and scientific ("why") approach, based on *practically proven* solid theory — model — measurements. The main focus will be to ensure that a practitioner can use the recommended step-by-step approach to improve the results of her daily activities.

TARGET AUDIENCE

- Primary audience will be pharmaceutical personnel, from low level R&D and production technicians to team leaders and department heads. Some topics will also definitely be of interest to people working in nutraceutical and generic manufacturing companies. The series will also be useful for academia and regulatory agencies.
- Each book in the series will target a specific audience, and since the format will be short, the price will be affordable not only for major pharmaceutical libraries but also for thousands of practitioners.

Welcome to the brave new world of practical guides to pharmaceutical technology!

Series Editor Michael Levin, PhD
Milev, LLC Pharmaceutical Technology Consulting

CHAPTER 1

Introduction

1.1 INTRODUCTION

The batch fluid bed granulation process is a well-established unit operation in the pharmaceutical industry; however, other process industries, such as food, nutraceutical, agro-chemical, dyestuffs, and other chemical industries, have adopted a fluid bed granulation process to address particle agglomeration, dust containment, ease of material handling, and modification of particle properties to provide flowability, dispersibility, or solubility to products, among other product enhancements.

Earlier, fluid bed was used as an efficient dryer, compared to other methods to dry a product. With the advent of newer technologies and drug delivery techniques, these units besides drying, are now routinely used for granulation, particle coating, palletization, and melt agglomeration to produce granules or modified release particles or pellets. Because of this versatility, these units are normally classified as multiprocessor fluid bed units.

The continuous fluid bed processing for drying and in some cases, agglomerating, has been used for certain high volume products. In subsequent chapters in this book, an attempt is being made to help the reader solve day-to-day challenges working with the fluid bed process.

Fluidization is the unit operation by which fine solids are transformed into a fluid-like state through contact with a gas (normally processes air). At certain gas velocities, the gas will support the particles, giving them freedom of mobility without entrainment. Such a fluidized bed resembles a vigorously boiling fluid, with solid particles undergoing extremely turbulent motion, which increases with gas velocity. The smooth fluidization of gas-solid particles is the result of equilibrium between the hydrodynamic, gravitational and inter-particle forces. The air stream requirements for minimum fluidization velocity vary depending on particle size, density, shape and even the surface properties.

How to Optimize Fluid Bed Processing Technology. DOI: http://dx.doi.org/10.1016/B978-0-12-804727-9.00001-6

Suspension and movement of particles in an airstream maximize the exposure of particle surfaces to air or gas, producing efficient evaporation. The primary factor influencing a fluidized-bed process is air flow. To understand and manipulate processing in a fluid bed, it is important to learn how airflow is generated, conditioned, and distributed through the bed during drying, agglomerating, and coating. We will discuss that in the next chapter.

1.2 ADVANTAGES AND CHALLENGES OF FLUIDIZED BED GRANULATION

1.2.1 Granulation

Most solid products in the pharmaceutical industry are manufactured using the wet granulation process. Fluid bed granulation is an established choice for improving the processing properties of pharmaceutical powders, such as flow characteristics and tablet compaction.

The fluid bed granulation can be classified as a *single pot process*, as the powders can be mixed, granulated and dried in the same unit, facilitating product transfers and minimum cross-contamination. In addition, the fluidized bed enhances heat and mass transfer between the fluidizing air and solid particles, leading to uniform temperature distribution within the product bed and relatively short processing times. In comparison with high shear granulation, the size distribution of granules produced from the fluidized bed technique is often narrower, with the absence of oversized granules. This reduces the need for re-granulation and accelerates drying.

Fluidized bed granules have also been reported to be more porous, less dense and more compressible than granules produced from high shear wet granulation. The optimal particle size for fluidization ranges from 50 to 2000 µm. The average particle should be somewhere between 50 and about 5000 µ, in order to avoid excessive channeling and slugging. Very fine particles tend to lump together due to the cohesive forces related to the very large surface areas; accordingly these particles normally fluidize badly at gas flow rates where excessive elutriation can be avoided. For fine particles less than 50 µm and particles which are not amenable to fluidization when moistened,

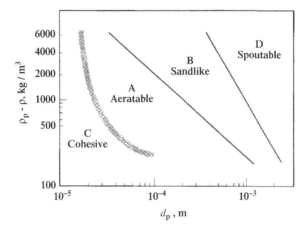

Figure 1.1 Geldart classifications of particles. From Geldart, D. Powder Technology. 7 (1973) 285–292.

mechanical means such as raking may have to be applied to the powder bed, increasing equipment, cleaning and maintenance costs. The critical size limit below which common pharmaceutical powders cannot be discretely processed is approximately 20 μm. Below this limit, steady fluidization without any retardation is difficult as indicated by Geldart's fluidization map. See Fig. 1.1.

Processing powder mixtures comprising components of vastly different densities poses yet another challenge, as disparities in fluidization behavior of the different formulation components may result in bed segregation and nonuniform mixing. During fluid bed granulation along with these powder properties, the spreading ability of the droplets of the binding liquid in the powder bed is also critical. As such, agglomeration in the fluidized bed process is highly dependent on the phenomenon of liquid spreading. Evidently, fluidized bed granulation is an intricate process and apart from material-related factors such as the nature and characteristics of the ingredients in the formulation, process factors related to the granulation and drying stages of the process also influence process outcomes.

1.2.2 Drying

Fluidized bed drying is a three-stage process, including a short preheating stage, a constant rate stage and a falling rate stage. The constant rate stage corresponds to a constant bed temperature. Rapid mixing of solids leads to nearly isothermal conditions throughout the fluidized bed, i.e., reliable control of the drying process can be achieved easily. The capacity of the air (gas) stream to absorb and carry away moisture determines the drying rate and establishes the duration of the drying cycle. Controlling this capacity is the key to controlling the drying process. The three elements essential to this control are inlet air temperature, dew point, and air flow. The higher the temperature of the drying air, the greater its vapor holding capacity.

1.2.3 Coating

Fluid bed (FB) coating offers the possibility to alter and to improve various characteristics of core particles such as the surface properties in a single unit operation. The challenges of using this technology are the difficulties in choosing the proper process conditions that lead to a constant coating quality and a robust process, especially during process up-scaling. The functions of coating are extensive, ranging from basic necessity to aesthetic purposes. Coating can be used to improve the chemical and physical properties of the substrate.

1.2.4 Challenges

Manufacturing a product with the fluid bed process can present numerous challenges. Challenges may be related to equipment, material properties, process parameters, material handling, potent compounds manufacturing, and process scale-up, etc. The following chapters provide the nature of the challenges encountered in each unit operation and possible solutions in detail. As mentioned earlier, the emphasis will not be on theoretical aspect but practical hands on solutions.

Table 1.1 is a list of possible challenges you may encounter during fluid bed processing. This is not an exhaustive list; your own material properties may pose different challenges because of interdependence of process variables, material variables and equipment variables. Understanding some of the interdependence of these variables will help you solve your specific challenges.

Table 1.1 A List of Possible Challenges During the Product Processing With Fluid Bed Process Technology and Possible Solutions will be Provided in the Subsequent Chapters

Possible Challenges

1. Uneven particle size distribution
2. Large lumps in the granulation
3. Bed stalling during the granulation
4. Nozzle clogging during granulation
5. Nozzle clogging during coating
6. Poor uneven fluidization during granulation
7. Poor fluidization during drying cohesive product
8. Poor product yield after the process is complete
9. Very fine granules
10. Poor process air temperature control
11. Low dose active pharmaceutical ingredient (API) distribution causing de-mixing
12. After granulation and drying product is desegregated
13. Particle agglomeration during coating
14. High shear granulated product has lumps after drying
15. Product sweating in the container after drying
16. Granules get localized wetting producing large lumps
17. Product has very high static charge and fluidization makes it worse
18. Granulating solution uses organic solvent and safety concerns
19. Filter keeps clogging
20. Particles agglomerate during Wurster coating
21. Process ingredients do not granulate for product with low angle of contact
22. The product cannot be scaled up
23. Product transfer in an integrated installation creates large lumps
24. Product is developed in one type of FB unit and needs to be scaled up in a different brand of commercial unit.
25. Process filter cartridges clog up due to sticky product being dried
26. Granulate product in fluid bed, which is high dose, poorly water soluble, low density, micronized API, which was successfully produced in high shear granulator
27. Process end point is not consistent from batch to batch
28. Best end point determination tool for granulation, drying and coating
29. Minimizing product breakage during drying
30. Transferring tray drying process to fluid bed drying
31. How to dry high shear granulated product with organic solvent without triggering Lower Explosion Limit threshold?

CHAPTER 2

Fluidization Theory

A fluidized bed is characterized by rapid particle movement, caused by the rising bubbles; consequently good particle mixing, high rates of heat transfer, and uniform temperature profiles are possible.

2.1 HOW DOES FLUIDIZATION TAKE PLACE?

The principle of operation of fluidized systems is based on the fact that if a gas is allowed to flow through a bed of particulate solids at velocity greater than the settling velocity of the particles and less than the terminal velocity for pneumatic conveying and equal to the minimum velocity of fluidization (Umf), the solids get partially suspended in the stream of upward moving gas which imposes high enough drag force to overcome the downward force of gravity. The drag force is a frictional force imposed by the gas on the particle; the particle imposes an equal and opposite drag force on the gas.

As the air flow rate is increased, the viscous drag on the individual particles in the packed bed increases, resulting in an increase in pressure drop (ΔP) across the bed. A point is reached where the drag force on individual particles becomes equal to their apparent weight; the bed then begins to expand in volume. Individual particles are now no longer in contact with adjacent particles, but are supported by the fluid. The bed is said to have just fluidized. For a very cohesive powder, the primary particles may have been held by van der Waals forces and may fluidize as agglomerated particles.

Thus as a particle becomes more fluidized, it affects the local gas velocity around it due to these drag forces. The influence of the drag force is more significant for irregularly shaped particles. Beyond the minimum fluidization point, any additional gas introduced should travel through the bed as bubbles. Fluidization behavior is a summation of various interactions and inter-particle forces. The van der Waals forces have been established to be dominant during powder handling and

How to Optimize Fluid Bed Processing Technology. DOI: http://dx.doi.org/10.1016/B978-0-12-804727-9.00002-8

fluidization, but the electrostatic forces also have a great influence on the behavior of the process. Other potential forces are liquid and solid bridges. The interactions, in which the inter-particle forces may appear, are particle-particle, particle-chamber, and particle-gas interactions. Two approaches, minimum fluidization velocity Umf and Geldart classification, are generally accepted as having the capability to predict and characterize the fluidization behavior of the solids [1] (Fig. 2.1).

The relationship between the air velocity and the pressure drop is as shown in Fig. 2.2 [2].

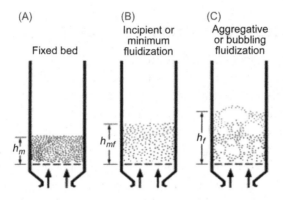

Figure 2.1 The behavior of the bed, as the air velocity is increased. (A) Gas or liquid (low velocity). (B) Gas or liquid. (C) Gas. From Kunii D, Levenspiel O. Fluidization engineering, 2nd ed. Stoneham: Butterworth-Heinemann; 1991 [1].

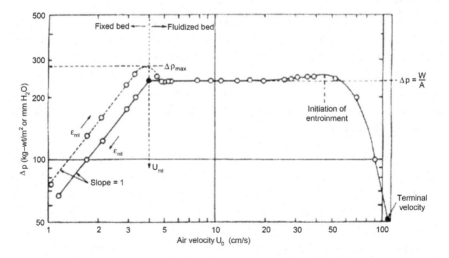

Figure 2.2 Typical pressure drop curve as a function of gas velocity. Adapted from Gomezplata A, Kugelman AM. Processing systems Chapter 4, in "gas−solid" handling in the process industries. In: Marchello JM, Gomezplata A, editors. Chemical processing and engineering, vol. 8. New York: Marcel Dekker, Inc.; 1976 [2].

2.2 UNDERSTANDING THE PARTICLES

In 1973, Geldart classified powders into four groups by the solid and gas density difference and mean particle size and their fluidization properties at ambient conditions. The Geldart Classification of Powders is now used widely in all fields of powder technology. Geldart classification consists of four groups that are characterized by the density difference between solid and gas, and the mean particle size of the solid phase. Understanding the nature of your product to be fluid bed processed is very critical. Geldart's classification of various particles is helpful and is provided in an abbreviated form in the Table 2.1.

2.3 TYPES OF FLUID BEDS

During the fluid bed process various fluid bed patterns can be observed depending on the fluidization velocity, product density, shape, and quantity of product in the bowl. The density, directly alters the net gravitational force acting on the particle, and hence the minimum drag force, or velocity, needed to lift a particle. The shape alters not only the relationship between the drag force and velocity, but also the packing properties of the fixed bed and the associated void spaces and velocity of fluid through them (Fig. 2.3).

• The gas velocity *(Umf)* calculated over the whole bed cross-section is called the *minimum or incipient fluidization velocity*. At the point of incipient fluidization, the bed takes on the appearance of a liquid; and it is self-leveling, will flow, and transmits hydrostatic forces (an object of lower density will float on the bed surface). At low gas velocities the bed of particles is practically a packed bed, and the pressure drop is proportional to the superficial velocity. As the gas velocity is increased, a point is reached at which the bed behavior changes from fixed particles to suspended particles. At the incipient point of fluidization, the pressure drop of the bed will be very close to the weight of the particles divided by the cross- sectional area of the bed. During the *incipient fluidization*, the particles are very close and not truly moving; for creating a homogeneous mixer of components, violent mixing is required which can be achieved by increasing the gas velocity.
• At gas flow rates above the point of minimum fluidization, a fluidized bed appears much like a vigorously boiling liquid; bubbles of

Table 2.1 Geldart's Classification of Powders (Umf = Minimum Fluidization Velocity and Umb = Minimum Bubbling Velocity)	
Group C powders	• Cohesive powders • They are difficult to fluidize, and channeling may occur. • Inter-particle forces greatly affect the fluidization behavior of these powders • Mechanical powder compaction, prior to fluidization, greatly affects the fluidization behavior of the powder, even after the powder had been fully fluidized for a while • Saturating the fluidization air with humidity reduces the formation of agglomerates due to static charges and greatly improves the fluidization quality. The water molecules adsorbed on the particle surface presumably reduce the van der Waals forces. • dp ~ 0–30 μm • *Example: flour, cement* *group C are difficult to fluidize*
Group A powders	• Size reduced by either using a wider particle size distribution or • Powders are easily aeratable • Characterized by a small Δp • Umb is significantly larger than Umf • Large bed expansion takes place before bubbling starts • Gross circulation of powder even if only a few bubbles are present • Large gas back mixing in the emulsion phase • Rate at which gas is exchanged between the bubbles and the emulsion is high • Bubble reducing the average particle diameter • There is a maximum bubble size • Δp ~ 30–100 μm • *Examples: flour* *powders in group A exhibit dense phase expansion after minimum fluidization and prior to the commencement of bubbling*
Group B powders	• Bubbling • Umb and Umf are almost identical • Solids recirculation rates are smaller • Less gas back mixing in the emulsion phase • Rate at which gas is exchanged between bubbles and emulsion is smaller • Bubbles size is almost independent of the mean particle diameter and the width of the particle size distribution • No observable maximum bubble size • Δp ~ 100–1000 μm • *Example: sand* *group B bubble at the minimum fluidization velocity*
Group D powders	• Spoutable • Either very large or very dense particles • Bubbles coalesce rapidly and flow to large size • Bubbles rise more slowly than the rest of the gas percolating through the emulsion • Dense phase has a low voidage • Δp ~ >1000 mm • *Examples: Coffee beans, wheat, lead shot* *in group D can form stable spouted beds.*

gas rise rapidly and burst on the surface. The bubbles form very near the bottom of the bed, very close to the distributor plate, and as a result the design of the distributor plate has a significant effect on fluidized bed characteristics. Increasing the superficial velocity above the minimum fluidization velocity results in the formation of "bubbles" which rise through the bed. The bed expansion results

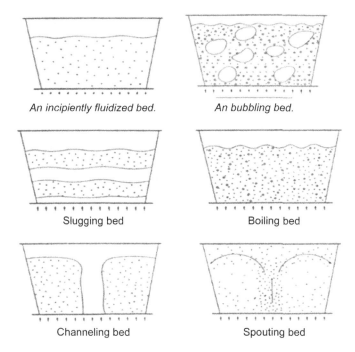

Figure 2.3 Types of fluid beds. From Parikh DM. Airflow in Batch fluid-bed processing. Pharm Technol 1991;15(3):100−10 [3].

mainly from the space occupied by the gas bubbles significantly with superficial gas velocity. These small bubbles tend to join together or coalesce as they rise through the bed. This creates larger and fewer bubbles than those near the distributor plate. In bubbling bed, mixing is caused not only by the vertical movement of bubbles and bursting of bubbles at the bed surface, but also by the lateral motion of bubbles as a result of the interaction and coalescence of neighboring bubbles.

- When the concentration of solids is not uniform throughout the bed, and if the concentration keeps fluctuating with time, the fluidization is called *aggregative fluidization.*
- A *slugging bed* is a fluid bed in which the gas bubbles occupy entire cross-sections of the product container and divide the bed into layers.
- A *boiling bed* is a fluid bed in which the gas bubbles are approximately the same size of the solid particles.
- A *channeling bed is* a fluid bed in which the gas forms channels in the bed through which most of the air passes.

- A *spouting bed* is a fluid bed in which the gas forms a single opening through which some particles flow and fall on the outside.
- As the velocity of the gas is further increased, the bed continues to expand and its height increases, whereas the concentration of particles per unit volume of the bed decreases. At a certain velocity of the fluidizing medium, known as entrainment velocity, particles are carried over by the gas. This phenomenon is called entrainment.

2.4 CONTROLLING GAS VELOCITY

Control of gas velocity is critical to effectively create a fluidized bed either for drying, agglomeration or particle coating. The fluidized bed advantage of fast heat and mass transfer can only be achieved if the particles are suspended in the gas stream during the processing.

To obtain a proper fluidization of your product, the following must be taken into consideration:

a. Quantity of the product (batch size)
b. Particle size, shape, and density
c. Flow properties of the powder
d. The capacity of the fluid bed and its capacity and location of the fan in relation to the location of the unit
e. Minimum and maximum recommended capacity of the bowl

The control of the velocity of gas can be achieved, first and foremost, by the air distributor that is selected. The selection of the distributor depends on the type of material and its particle size, density, shape, quantity, the fan capacity and location of the system, among other considerations. Selection and further description for the distributor plates are provided in Chapter 3. The distributor type and geometry has a significant effect on the value of minimum fluidization velocity, where increasing the hole pitch of a perforated plate distributor decreases *Umf.*

2.5 ADDRESSING FLUIDIZATION CHALLENGES

- Determination of minimum fluidization velocity and bed expansion:
  i. Note the weight of product and their particle size distribution.
  ii. Supply incoming process air with a low flow rate (the lowest measurable) to the bed and note the flow rate. Also note the bed height.

iii. Increase the gas flow slightly and again note the flow rate, pressure drop and the bed height.

iv. Repeat (iii) until the maximum flow rate is reached. Observe the top surface and sides of the bed and note the flow rate when the bed just becomes fluidized. (Particles begin to vibrate).

v. Repeat (iii) starting from a high flow rate and decrease the flow rate slightly in steps. Note the flow rate when the bed is just de-fluidized.

- If you observe erratic fluidization or no fluidization of product in the product bowl, this could be the result of:

  i. Differences in particle sizes and densities of the ingredients being granulated creating differences in the velocity that lighter particles experience compared to denser particles. By selecting similar particle sizes and densities may help creating uniform fluidization.

  ii. The quantity of product in the bowl is too little and the fluidization velocity and/or air volume is high.

  iii. The air distributor may be inappropriate providing the high velocity for incoming air, by changing the more open air distributor plates will provide lower velocity and could help fluidize the product.

  iv. If the fluidization is not occurring at all this could be several reasons such as

    - The quantity of product in the bowl is more than the capacity of the bowl and the blower(fan) or,
    - the density of product may be high,
    - the air distributor could be more open and the fluidization velocity is not high enough. Increasing the incoming air volume will provide kinetic energy needed to fluidize the batch.
    - In extreme cases, materials such as most cohesive powders or wet granulated product, or centrifuge cake being dried, a mechanical rake at the bottom of the bowl or inside the lower plenum is provided by some manufacturers of the fluid bed units to facilitate breakage of lumps and thus helping fluidization.
    - Particles smaller than 50 microns are difficult to fluidize due to its cohesive nature. These are the particles belong to the group C in Geldart's classification of powders. These particles tend to form agglomerates of random size and shape because of high ratio between the surface and the volume between the particles. This strong interaction between the powders affects

fluidization. Mechanical rake can assist in this case to modify the hydrodynamic behavior of powders and assist fluidization.
- Gas velocity that correspondence to the maximum bed pressure drop increases with increase of the open fraction of the air distributor.
- During drying, uneven fluidization can be noticed by a very short drying time and a low pressure drop across the bed.

v. Determining the fluidization volume that will be optimum for the process consideration:
- for example, for fluid bed granulation, the powder flow pattern should be like a fountain or a free downward flow, so the particles in their upward movement get the atomized binder solution droplets on them. A higher than optimum fluidization would create powders to entrain and occlude the filters.
- exhaust air temperature can be used to detect poor fluidization. Rise of rapid exhaust air temperature may indicate improper fluidization [4].

2.6 SUMMARY

Fluidization in a fluid bed process is critical for creating homogeneous mixer of particles. Whether agglomerating, drying, or coating, fluidization at a proper level within the processor is required. Uneven fluidization will result in variation in the moisture content, particle size distribution or un-granulated product resulting in poor content uniformity. Establishing the minimum fluidization velocity and controlling the bed by observing pressure drop across the bed and the filters will provide better control of the process.

REFERENCES

[1] Kunii D, Levenspiel O. Fluidization engineering. 2nd ed. Stoneham: Butterworth-Heinemann; 1991.

[2] Gomezplata A, Kugelman AM. Processing systems Chapter 4, in "gas–solid" handling in the process industries. In: Marchello JM, Gomezplata A, editors. Chemical processing and engineering, vol. 8. New York: Marcel Dekker, Inc.; 1976.

[3] Parikh DM. Airflow in Batch fluid-bed processing. Pharm Technol 1991;15(3):100–10.

[4] Gao JZH, et al. Importance of inlet air velocity in fluid bed drying of a granulation prepared by high shear granulator. AAPS PharmSciTech 2000;1(4) TechNote.

CHAPTER 3

Fluid Bed Processor Equipment and Its Functionality

3.1 INTRODUCTION

As stated in an earlier chapter, understanding the process equipment will help in designing the process, and in turn will be able to optimize the fluid bed process. The fluid bed equipment is normally installed in an good manufacturing practice (GMP) area and the supporting systems such as air handling units (AHUs), duct work, fan or blower, are all either in a technical area behind the processor unit, or above on the mezzanine or outside of the building. Knowing the functionality of each component of the fluid bed system and your process understanding and requirements, equipment should be selected and installed accordingly.

Fluid bed systems vary in function and purpose. Because of the versatility of the processor, it is normally designated as a "multi-processor." Different modules are used to carry out different unit operations such as drying, agglomeration, particle coating, pellet formation and coating, melt agglomeration, and taste masking. Fig. 3.1 shows typical modules that are used to carry out various unit operations by changing the product container module.

Selecting a fluid bed processor requires knowledge of several factors. Early in the product development stage, selection of fluid bed technology is determined as a result of laboratory experiments. Based on the initial process development, one has to understand requirements for the pilot size equipment and eventually commercial manufacturing for the given product.

While considering the commercial scale unit, careful understanding of scale-up issues and material handling options should be evaluated. At the end of the day, process optimization depends on how efficiently commercial production is executed with minimal manpower and time.

How to Optimize Fluid Bed Processing Technology. DOI: http://dx.doi.org/10.1016/B978-0-12-804727-9.00003-X

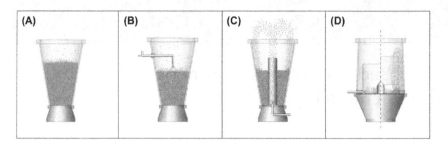

Figure 3.1 Various modules for fluid bed processing. (A) Drying. (B) Agglomeration. (C) Coating or granulation. (D) Rotary fluid bed for palletization/coating. Courtesy: GEA Pharma systems.

It is advisable to meet with equipment manufacturers and establish a user requirement list based on your process. The engineering staff from the supplier of equipment and your engineering department can specify all the relevant utility, installation and other support services requirements. If you are not familiar with this technology it is very useful to visit a few manufacturing facilities using fluid bed processor for various applications. Some of the questions that need to be answered are listed below. *(This is not necessarily an all-inclusive list, and individual product and process requirements may raise additional questions.)*

3.2 QUESTIONS BEFORE SPECIFYING THE EQUIPMENT

A well thought out User Required Specifications could be prepared from the questions listed below.

Commercial/pilot Fluid Bed Equipment Specifications

- What is the unit operation you want to use fluid bed for? (e.g., drying, agglomeration, coating)
- What is the product that needs to be processed? Is the product a 100% drug substance or does it have a composition of drug substance with other excipients?
- What are the individual physico-chemical properties of API and excipients?
- What is the density, moisture content, flow behavior of both API and excipients?
- What are the product characteristics as a mixture of API and excipients?
- What are the physico-chemical characteristics of excipients?
- What is the percentage of drug to excipient ratio?

- What is the solubility of drug substance and other excipients?
- Is it a potent drug substance? What is the potency classification? Does the system have to be contained? Do we need a glove chamber?
- How expensive is the drug substance?
- Is the material toxic, flammable, corrosive or abrasive?
- What is the particle size distribution required in the final granulation?
- How will the fluid bed processed product be utilized? Will the end product be granules only or will it be combined with other ingredients for further processing or coated for modified release application or be tableted, encapsulated? In short, what is the end use for the product?
- Plant: How much space is available in the plant? How warm, humid and clean is the plant air? What fuel and power sources are available?
- In what geographical area will the unit be installed?

3.2.1 Granulating in Fluid Bed
- In case of granulating in fluid bed, what binder will be used, and is the binder in solution or added dry? What is the binder solution concentration and viscosity? Can you pump the binder solution? What is the solvent (is it water, organic solvent or mixture of organic solvent and water?)
- What are the fluidization characteristics of the product to be granulated?
- Does the material need to be processed as a batch or continuous?
- Do we need humidification and dehumidification of the incoming air? (This is generally highly recommended to assure consistent process.)
- What will be the batch size and what is are the current and projected requirements for the final product?

3.2.2 Drying in Fluid Bed
- In case drying of granulation only, determine how the wet mass from the mixer will be added to the fluid bed bowl. Manually, in an integrated equipment set up?
- How cohesive is the wet mass? What is the moisture/solvent content of the wet mass?
- Is the wet mass very cohesive and difficult to fluidize without additional assistance? Do we need a mechanical rake to break up lumps, to facilitate fluidization?

- In case of drying granulated product, discharging from the high shear granulator can also be as simple as "dumping" the product into a waiting fluid bed container if the binder solvent is water.
- Product containing organic solvent, wet mass can be vacuum transferred to a running fluid bed. (*Since the running fluid bed is under negative pressure, it acts like a vacuum cleaner and vacuums the product in the fluid bed from the high shear mixer or through the integrated system. While not part of the actual process of fluidizing, charging and discharging are integral to the machine's safe operation and may affect product quality, batch yields and integrity if not properly considered and carried out*).
- If the wet mass contains organic solvent, is the transfer required to be gradual so as not to exceed the lower explosion limit during drying?
- Is the material's moisture bound (chemically trapped inside the particles), unbound (not attached to the particles, also called free moisture) or both? What are the material's initial and final moisture contents? What are the maximum permissible drying temperature and probable drying time for the material?
- Final product quality requirements: Can the material shrink, degrade, overdry or become contaminated during drying? How uniform must its final moisture content be? What should the final product temperature and bulk density be?
- Must the capacity of high shear mixer and the fluid bed capacity be the same? Or is there possibility that two granulation loads will be used for drying cycle? (*So the capacity of the granulator will determine the quantity of product that needs to be dried and hence the size of the fluid bed unit that would be required*)
- Will the equipment be dedicated for only one product?

3.2.3 Coating in Fluid Bed
- What is the particle size of the substrate to be coated?
- What is the coating solvent?
- Is it functional coating only? Or solution/suspension layering?
- Batch size versus the coating chamber capacity

3.3 FLUID BED PROCESSOR COMPONENTS

The fluid bed processor contains following main components:

- Air handling unit (AHU)
- Machine tower with lower plenum, expansion chamber & filter housing

Figure 3.2 Typical fluid bed processor set up integrated with solution delivery system and discharge through the mill in the IBC. Courtesy: IMA, Italy.

- Product container (either portable or swing-out type) with distributor plate
- Inlet and exhaust air duct with temperature probe
- Fan or blower
- Spray nozzle (for granulation)
- Solution delivery system (tank, pumps, tubing, etc.) for granulation
- Control panel

   Fluid bed dryer and fluid bed processor (granulator/coater) has the following common components as illustrated in the Fig. 3.2.

3.3.1 Machine Tower

The machine tower comprises ductwork bringing the conditioned air from the lower plenum, through the product container and the product, passing through the expansion chamber, process filters and to the exhaust carrying the moisture.

3.3.2 Air Handling Unit (AHU)

Use of an AHU allows filtering, heating, cooling, and humidity removal of the inlet process air. The air dehumidification is especially important when the production unit is located in a climate with large moisture variations, as the binder liquid evaporation rate is determined by the processes of heat and mass transfer. If needed, re-humidification is commonly accomplished by injecting clean steam into the airstream following the heating and cooling stages. Outside air is filtered and cooled by either a direct expansion coil (refrigerant coil) or a chilled fluid coil. The chilled coil also serves to dehumidify the air. As moist air travels across a cooled surface, condensation occurs. Typically, the minimum dew point that is attained by chilled coils is +36°F (+2°C). If drier air is required, a desiccant dehumidifier is used. Air passes through a sorbent material (silica gel or alumina) in a rotating wheel and a portion of the wheel is regenerated by back purging with heated air. The injection of dry steam may also be used to increase the humidity to precise levels. Heating coils are used to heat the process air to prescribed conditions. Finally, the process air is delivered though a HEPA (High Efficiency Particulate Arresting) filter to produce pharmaceutical grade air.

3.3.3 Fan

The fan is located at the downstream end of the unit which creates suction to pull the air from the AHU, and keeps the system at a lower pressure than the surrounding atmosphere. Once the air leaves the exhaust filters, it travels to the fan and out. In case of solvent process, a catalytic oxidizer is installed to burn the solvent vapors.

Filtration of the air exiting the fluid bed after the filter chamber is usually done to prevent contamination of the local environment. Such systems can involve:

- Dust collectors
- Scrubbers or incinerators (thermal oxidizers)
- HEPA systems with bag-in, bag-out feature for operator exposure protection for potent compounds
- Closed loop recovery systems (designed to recycle and/or capture the fluids and products for disposal or possible re-use)

3.3.4 Product Container

The product container holds the product for processing. The process air is introduced evenly at the bottom of the product container through an inlet air plenum. If the air is not properly distributed before it reaches the bottom of the container, erratic fluidization can occur. The container volume should be chosen such that the container is filled to at least 35%−40% of its total volume and no more than 90% of its total volume. These distributors are made of stainless steel and are available with a 2%−30% open area. The open area determines air velocity at a given air volume. Typically, the distributor should be chosen so that the pressure drop across the product bed and air distributor is 200−300 mm of water column. Most common air distributors are covered with a 60−325 mesh fine screen to retain the product in the container. Keeping the screen and air distributors clean is challenging. Partially to address the cleaning problems and partially to provide efficient processing, a number of manufacturers have introduced air distributors that eliminate the use of a retaining screen.

3.3.5 Solution Delivery System

Several types of spray nozzle systems are available for use in the fluidized bed processor. The two-fluid nozzle system is most popular as it is capable of functioning at very slow liquid flow rates and allows the control of droplet size independently of flow rate. A two-fluid nozzle is used where binder liquid is pumped through an orifice and atomization air facilitates the breakup of the liquid stream into fine droplets. The higher the pressure, the finer the droplets. A single-headed nozzle is generally adequate for the pilot size units up to 100 kg batch size. Depending on the size of the units, three single-port nozzles, or a three-port or six- port nozzle can be used as you scale the process up. See Fig. 3.3.

For solution delivery normally a peristaltic pump is used for transferring the binder solution to the spray nozzle. One pump per nozzle is highly recommended.

An oil- and moisture-free compressed air source which can show the volume and the atomization pressure gauges (indicators) to atomize the solution is required.

3.3.6 Utilities

The utilities and space involved need to be discussed prior to purchasing the equipment. Obviously, the actual equipment size is but a small part of the entire scope of work necessary to implement a larger system. The unit and its support equipment must be able to be installed in current building facilities.

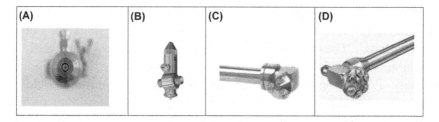

Figure 3.3 Two fluid nozzles used for fluid bed granulation. (A) Two fluid nozzle showing annular opening for compressed air surrounding the liquid port. (B) Single port nozzle. (C) Three port nozzle. (D) Six port nozzle. Courtesy: The Glatt Group.

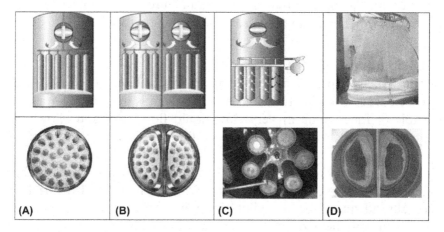

Figure 3.4 Different types of process filters. (A) Single filter bag. (B) Split filter bags. (C) Cartridge filters. (D) Bonnet with split sections (filter for modified release polymer coating). (A, B, C) Courtesy: The Glatt Group.

3.3.7 Process Filters

During the granulation or drying process, filter bags with socks are mechanically shaken to dislodge any adhered product, while a cartridge uses a low-pressure compressed air blowback system to do the same. Fig. 3.4 shows different filter arrangements.

Filter bags are made from polyester, nylon or Polytetrafluroethylene (PTFE) material. Typically, a supplier will supply filters, which are made up of polyester or nylon material with carbon filament embedded in it. The micron opening is normally 20 microns unless you have a product that requires a smaller micron opening and the system has a capacity (mainly the capacity of the blower) to overcome the pressure drop due to a smaller filter opening. A filter with an opening as low as 1 micron is available. Selection of the filter type and its micron opening is very critical. If the fluid bed processor is exclusively used for drying the product, then a single filter bag set (see Fig. 3.4 A) should be adequate and is less expensive. However, keep in mind that this may have an effect on drying time consistency and the resulting dried granule particle size distribution, because every time the filter shakes, the bed will collapse during the duration and frequency of the shake time. Hence it is prudent to select a split filter bag option (see Fig. 3.4 B). These split bag designs shake alternatively without process interruptions.

Cartridge filters (see Fig. 3.4 C) use one or more cylindrical cartridges situated in a housing. Air with particulates is directed to the outer surface, then exhaust air flows through the cartridge to the hollow center and out, while particulates are trapped. These filters use media that are pleated, for increased surface area, usually with multiple, concentric pleated sheets wrapped around a central core. The sheets can be made from any thin material that can be folded without cracking, including cellulose, polyester, fiberglass, PTFE, and polyethersulfone.

In the case of potent compounds, a stainless steel cartridge installation should be considered. This can only be justified if you have a high value, high potency product and operator exposure to the compound is deemed hazardous. The cost of stainless steel filters is very high. However, unlike cloth or other cartridge filters, stainless steel cartridges can be set up to clean in place thereby making cleaning validation easy.

For a functional coating of pellets or particles where no filter shaking is desired, a split design bonnet shown in Fig. 3.4D is used.

3.3.8 Control Panel

A fluid bed granulation process can be controlled by pneumatic analog control devices, or state of the art, programmable logic controllers or computers. The electronic-based control system offers not only reproducible batches according to the recipe but a complete record and printout of all the process conditions. Process control technology has changed very rapidly and it will continue to change as advances in computer technology take place and as the cost of control systems falls. The CFR Part 11 requirements mandated by the US FDA has created a number of approaches to assure these control systems are complying with the current regulations.

3.4 SUMMARY

Understanding the process equipment and its functionality is critical in designing the process and obtaining optimized results. Adequate laboratory experimentation and design of the process to obtain desired product attributes should be the starting point before selecting and specifying equipment. User required specifications will help procure the right equipment and supporting systems.

CHAPTER 4

Process Development

4.1 PROCESS SELECTION

Optimization of the process starts with the successful development of a robust process. The Design of Experiments (DoE) does offer a learning tool toward that objective. With DoE critical process variables can be identified. Understanding the responses from this experimentation, one can define operating limits of the process as it gets scaled up. Process variables found to be the most significant during the DoE may still apply to the pilot scale and ultimately for the commercial batches.

A process development starts with all of the ingredients of the formulation being compatible and when the potency is maintained of the active ingredient, during the preformulation studies.

Once the formulation is considered to be acceptable, then the process selection will require consideration of the critical quality attributes (CQA) of the dosage form being considered. The selection of fluid bed granulation or high shear granulation will depend on the physicochemical properties of the ingredient composition and quantities of each ingredient and their solubility, particle size, density, flow properties, stability, etc. Consistency of the ingredient physical and chemical specifications is of paramount important for optimization of the granulation process. Variations in the physical properties of the powders to be granulated can give rise to processing problems. An investigation and understanding of the interactions between inactive ingredients and active ingredient(s) is very important. The function of each ingredient in the matrix should be understood and controlled in a manner that tolerates variations and ranges that make for a rugged product and not sensitive to small perturbations in content or process. Availability of the technology during the development stage will have influence on the selection of the process as well. As is well known, the granulation from a high shear mixer will be denser with a low porosity while the

How to Optimize Fluid Bed Processing Technology. DOI: http://dx.doi.org/10.1016/B978-0-12-804727-9.00004-1

granulation produced using a fluid bed will produce porous granules with a good flow and compression properties. The fluid bed granulation process is normally of longer duration than high shear granulated and dried in the fluid bed process. Fluid bed processing requires accurate and reliable control of all the process parameters. Important parameters that must be controlled for wet granulation are product temperature, atomization air pressure, air dew point or humidity, and spray rate of the binder solution. Physical properties of fluidized bed granulations are strongly influenced by the binder and its concentration. An increase in the binder content of the formulation increases the binder adhesiveness, forming granules of low friability and larger average size.

4.2 QUALITY BY DESIGN

Quality by design (QbD) is a concept that incorporates some of the following principles:

- All critical sources of variability are identified and explained
- Variability is understood and appropriately managed during the manufacturing process
- Product quality attributes can be accurately and reliably predicted over the design space
- Critical design principles incorporate these three actions—evaluate, plan, and measure.

The use of QbD principles during product development provides opportunities to facilitate innovation and continual improvement throughout a product's lifecycle. A dosage form created using QbD principles should demonstrate a ruggedness or reproducibility factor that tolerates acceptable, understood and controlled variation. Formal experimental designs can be used to identify critical and interacting variables that might be important for constructing the design space and ensuring the quality of the drug product. The ICH Q8 guideline (2005) states that the unexpected results of pharmaceutical development studies can be useful in constructing the design space.

Fig. 4.1 summarizes the QbD approach for development.

Process development approach

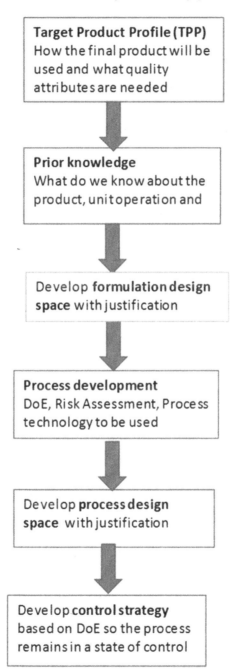

Figure 4.1 QbD approach for process development.

4.3 PROCESSING IN FLUID BED

Major processing factors affecting granule formation and growth in fluid bed granulation are:

- Temperature and volume of the fluidizing air during granulation cycle
- Height of the spray nozzle with respect to fluidized solids
- Rate of binder addition
- Degree of atomization of the binder solution/dispersion

4.3.1 Establishing the Critical Process Parameters

The fluid bed process has multiple variables and they interact with each other. It is essential to determine the critical process parameters. Knowing the effects of these parameters, including their interactions, is critical to control them. Process optimization depends on the best set of critical and controllable variables to produce a product with desired attributes all the time. Variation in process parameters and thereby in CQA can elicit critical variation to the end product. To understand the impact of process parameters a "design space" is required within which robustness of the process parameters can be demonstrated.

In the case of fluid bed granulation, knowing the desired granulation properties or (CQA) of the granules is a good starting point to determine what process parameters will affect those attributes. By use of the Ishikawa diagram, critical process parameters with impact on the desired granule properties such as moisture content, particle size distribution, and flowability can be determined and risk analysis of the fluid bed granulation can be performed. A typical Ishikawa diagram can be prepared for a fluid bed granulation process as shown in Fig. 4.2. The critical variables can then be ranked qualitatively based on prior knowledge and initial experimental findings. A similar Ishikawa diagram can be prepared for other fluid bed operations such as drying or coating, as well as the entire solid dosage manufacturing process steps to determine the critical parameters that need to be identified and controlled to produce the Target Product Profile (TPP).

Once the critical variables are established for the process, to create the design space, DoE can be performed to evaluate the impact of the higher ranked variables, and to gain greater understanding of the

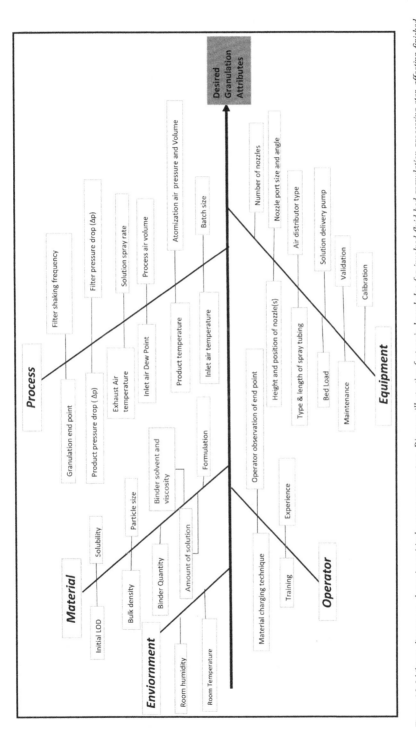

Figure 4.2 Ishikawa diagram to determine critical process parameters: Diagram illustrating factors in knowledge for involved fluid bed granulation processing step affecting finished product Critical Quality Attributes (CQAs) of final granulation.

process. This also helps to develop a proper control strategy. A risk assessment and formal experimental designs can lead to an understanding of the linkage and effect of process inputs on product CQAs and also help to identify the variables and their ranges within which consistent quality can be achieved.

4.4 RISK ASSESSMENT AND MANAGEMENT

The assessment of criticality as it relates to drug product quality is typically determined as a function of risk. Delineating criticality is a function of assessing risk, predicated on understanding the relationship of process variables and material attributes to the quality attributes of the drug product.

After the selection of the process it is advisable to understand the risk for the fluid bed processing and predict potential difficulties with process control and/or vulnerability to various forms of contamination. Risk identification can be done by the knowledge space and matrix investigation of critical process variables affecting CQAs. When considering risk assessment of fluid bed processing, the list in Table 4.1 is a good starting point for a risk evaluation for the fluid bed granulation process.

4.4.1 Risk Reduction and Control

Risk reduction and mitigation can be achieved by implementation of a control strategy irrespective of scale on the basis of overall development by QbD. A control strategy should be designed to ensure that a product of required quality would be produced consistently by the proposed process without probability of failure at a larger scale. The

Table 4.1 Risk Assessment for Fluid Bed Granulation Process With Failure Mode		
Critical Process Parameter	Failure Mode (Critical Event)	Effect on CQAs With Respect to QTPP (Justification of Failure Mode)
Liquid spraying rate	Higher RPM	Produce larger granules (lump) = Disintegration and dissolution can be affected
Atomizing air pressure	Lower pressure	Unevenly distributes drug binder solution = Content uniformity can be affected
Product temperature	Very high Inlet/Product/Exhaust temperature	Rate of degradation may be affected = Impurity profile may be affected
Fluidizing air flow rate	Higher CFM	Attrition and evaporation produces fines by which process efficiency (%yield) can be impacted

elements of the control strategy should justify how in—process controls and the controls of input materials (drug substance and excipients), intermediates (in—process materials), the drug products container and closure system will contribute to the final product quality. These controls should be based on product, formulation and process under-standing and include, at a minimum, control of the critical process parameters and material attributes. Sources of variability that impact product quality should be identified, appropriately understood and subsequently controlled.

Summarizing, a control strategy should include the following:

- Control of input material attributes (e.g., drug substance, excipients, primary packaging materials) based on an understanding of their impact on processability or product quality;
- Product specification(s);
- Controls for unit operations that have an impact on downstream processing or product quality (e.g., the impact of drying on degradation, particle size distribution of the granulate on dissolution);
- In-process or real-time release testing in lieu of end-product testing (e.g., measurement and control of CQAs during processing);
- Monitoring program (e.g., full product testing at regular intervals).

4.5 PROCESS CONTROL

Fluid bed processing requires control of certain key factors that affect the process critically. These are:

- Proper functioning of the controller and recorder for the fluidizing air to ensure close temperature and dew point controls throughout the process.
- Minimizing variations in active drugs or excipients, (especially binder) due to different vendors or lot-to-lot variations from one vendor.
- In case of granulation, uniformity of binder spray rates throughout the granulating phase.
- Good maintenance procedures for the pump and spray nozzle assembly and exhaust filter assembly to avoid process disruptions.
- Establishing a preventative maintenance program for the entire system.

- Providing built-in process alarms and signals, e.g., filter bag rupture detection device, etc., to spot process problems instantly.
- Promoting of good operator vigilance and training and process "know-how" to detect process problems in time to allow corrective actions.

Process analytical technologies (PAT) helps ensure final product quality by designing, analyzing, and controlling manufacturing through timely measurements of critical quality and performance attributes of materials and processes. For the fluid bed processor, instrumentation on the machine includes inlet and exhaust temperature, product temperature, inlet air humidity (dew point), total air volume, pressure drop across the product and the filters, air atomization pressure, spray rate among other measurements. For in-line measurement of particle size growth during the granulation, PAT tools such as focused beam reflectance measurement and Parsum spatial filtering technique (SFT) are used. Data acquisition and analysis can be performed using Parsum-View software, which can display real-time graphs of particle size and distribution. The Parsum probe is based on SFT.

4.6 CASE STUDY

Rambali et al. optimized the granulation process by DoE approach [1]. Summary of this approach to optimize the process of fluid bed granulation is presented below:

A 30 kg batch of lactose and starch was granulated with solution of hydroxypropyl methyl cellulose in a fluid bed with a face-centered central composite design for the geometric mean granule size. First a Plackett–Burman design was applied to screen the inlet air temperature, the inlet flow rate, the spray rate, the nozzle air pressure, the nozzle spray diameter, and the nozzle position (Table 4.2). The Plackett–Burman design showed that the key process parameters were the inlet flow rate and the spray rate and probably also the inlet air temperature. Afterward a fractional factorial design (2^{5-2}) was applied to screen the remaining parameters plus the nozzle air cap position and the spraying time interval. The fractional factorial design showed

Table 4.2 Settings of the Granulation Process Parameters in the Plackett–Burman Screening Design (Ref. [1])

Process Parameters	Settings
Inlet air flow rate	140–286 m³/h
Nozzle spray diameter	1.2–2.2 mm
Nozzle height position	1.0–3.0
Nozzle air pressure	1.5–2.5 bar
Inlet air temperature	50–70°C
Spray rate	58.0–135.6 g/min
Note: Nozzle air cap position is fixed (position 5), spraying time is 35 s, and filter shaking time is 7 s.	

Table 4.3 Centered Two-Level Factorial Design of the Inlet Air Flow Rate and the Inlet Air Temperature at a Spray Rate of 58.0 g/min. (Ref. [1])

Process Parameters

Run	Inlet Air Flow Rate (m³)	Inlet Air Temp.°C	R = Granule yield Between 75 and 500 μm (%)	D50 μm = Geometric Mean Granule Size (μm)
1	286	40	93.6	322
2	286	70	88.8	235
3	213	55	94.1	334
4	140	40	45.8	697
5	140	70	95.4	288
Note: Nozzle spray diameter is 1.8 mm, nozzle pressure is 2.5 bar, nozzle height position is 2.				

that the nozzle air pressure was also important. The target values for the granule yield (between 75 and 500 μm) and the geometric mean granule size (between 300 and 500 μm) were reached during the screening experiments. These regression models were used to optimize the granulation process to obtain a granule size between 300 and 500 μm. Additional experiments confirmed that these models were valid. Other granule properties, namely the geometric standard deviation, the Hausner index, the angle of repose and the moisture content, were evaluated at the optimal operation conditions (Table 4.3).

A quarter fractional (2^{5-2}) factorial design with the parameters that were temporarily considered less important in the Plackett–Burman design, and on two additional parameters that at first were not investigated in the Plackett–Burman design. This new factorial design consisted of 12 runs. The central point was replicated four times for the experimental error to evaluate the main effects. The parameters that were most significant in the Plackett–Burman design (the spray rate, the inlet airflow rate, and the inlet air temperature) were fixed in this new design.

Authors concluded from the applied Plackett–Burman design and the quarter fractional factorial design that the key granulation process parameters are the spray rate, the inlet air flow rate, the inlet air temperature, and the nozzle air pressure. See Tables 4.4 and 4.5.

Authors also concluded that in this study the granule size of the granulation could be optimized empirically, by considering the process variables. Optimal granules were obtained at low and central levels of the spray rate and inlet air humidity and central or high levels of the inlet air temperature and airflow rate. An optimal granule size was obtained at low levels of the powder bed moisture content and at low and central levels of the droplet size and at central and high levels of the airflow rate.

4.7 SUMMARY

We now have tools available for increasing the process understanding and quality of finished products. PAT, risk analysis, systematic experiments (DoE) and design space will likely be important elements of R&D. To be able to develop a robust process that can be optimized on a production scale, a clear definition of CQA should be defined. Once that is understood, the formulation optimization is necessary including design space for the formulation. The selection of the process will depend on the CQA and the properties of API and the excipients. Once the process is selected, critical process parameters need to be identified, risk assessment and DoE performed to create the design space. Variables with the greatest impact on the final CQA had to be controlled by in-line or off-line tools. As the product moves from development to pilot plant to commercial production, the process needs to be maintained in a constant state of control. The robustness of the process thus starts at the R&D stage. The design space is dynamic and begins at the initial stage and continues to evolve over

Table 4.4 Plackett−Burman Screening Design of Process Parameters for the Granulation Process With Fluidized Bed (Ref. [1])

| | Process Parameters | | | | | | Response | |
Run	Air flow Rate (m³/h)	Nozzle Diameter (mm)	Nozzle Height Position	Nozzle Air Pressure (bar)	Inlet Air Temperature °C	Spray Rate (g/min)	R = Granule Yield Between 75 and 500 μm (%)	D50 μm = Geometric Mean Granule Size (μm)
1	140	1.2	3	2.5	70	77.5	81.6	470
2	140	2.2	3	1.5	70	77.5	59.0	602
3	286	2.2	3	1.5	70	135.6	47.4	694
4	286	2.2	1	2.5	70	77.5	90.3	298
5	140	1.2	1	1.5	50	77.5	0.0	>694[a]
6	140	1.2	1	2.5	70	135.6	0.0	>694[a]
7	140	2.2	1	1.5	50	135.6	0.0	>694[a]
8	286	1.2	3	1.5	50	77.5	84.7	457
9	140	2.2	3	2.5	50	135.6	0.0	>694[a]
10	286	1.2	1	1.5	70	135.6	64.8	562
11	286	1.2	3	2.5	50	135.6	2.9	>694
12	286	2.2	1	2.5	50	77.5	97.5	297

[a]Failed runs are set to d50 > 694 μm.

Table 4.5 Centered Quarter Fractional (2^{5-2}) Factorial Design (Ref. [1])							
	Process Parameters					Response	
Run	Nozzle Diam. (mm)	Nozzle Height Position	Nozzle Air Pressure (bar)	Nozzle Air Cap Position	Spraying Time Interval (s)	R = Granule Yield Between 75 and 500 μm (%)	D50 μm = Geometric Mean Granule Size (μm)
1	2.2	1	2.5	1	70	96.4	348
2	2.2	3	1.5	5	35	82.3	528
3	2.2	3	2.5	5	70	96.7	352
4	1.8	2	2	3	52.5	85.2	494
5	1.2	3	2.5	1	35	94.4	458
6	1.8	2	2	3	52.5	94.7	471
7	1.2	1	2.5	5	35	97.6	355
8	1.8	2	2	3	52.5	93.5	444
9	1.8	2	2	3	52.5	95.4	426
10	1.2	1	1.5	5	70	82.8	504
11	2.2	1	1.5	1	35	57.9	612
12	1.2	3	1.5	1	70	65.2	565

Note: Spray rate was 68 g/min, Inlet air flow rate was 213 m^3/h and inlet air temperature was 55° C.

the entire lifecycle of the process. Increased development resources may be required to achieve knowledge-based understanding of the process. The case study presented shows how a DoE can be conducted and evaluated.

REFERENCE

[1] Rambali B, Baert L, Massart DL. Using experimental design to optimize the process parameters in fluidized bed granulation on a semi-full scale. Int J Pharm 2001;220:149−60.

CHAPTER 5

Granulation

5.1 THEORY

5.1.1 Process

The fluid bed granulation process involves spraying a binder solution using two fluid nozzles on fluidized particles and agglomerating while concurrently evaporating the solvent. The success of the particle coalescence depends on the availability of the liquid at the particle surface and the strength of the liquid bridge between two particles to facilitate a successful coalescence. The particles are continuously mixed within the fluidized bed and the fluidizing gas provides the necessary heat for drying and carrying away the moisture. After the spraying is stopped, the remainder of the moisture is evaporated during the final stage of drying until the required moisture content is obtained. Fluid bed granules are characterized by a porous surface and interstitial void space. This results in an increased wicking of liquid into the granules and improved disintegration or dispersibility. In addition, because of these void spaces, the bulk density of fluid bed granules is lower than that attainable by other granulation techniques.

The volume of granulating liquid needed depends mainly on the solubility of the drug and/or excipients. Insoluble drugs require larger volumes of granulating liquid than soluble drug preparations. Particle size distribution, particle shape, surface roughness, and liquid, equipment and process characteristics all affect the amount of liquid required. Spraying and evaporation, these two conditions must be in perfect balance to have a functioning fluidized granulation process; if either of these limits is exceeded, liquid will accumulate in the bed causing the bed to collapse or product will not granulate. The spray rate of the binder solution is a critical factor in determining the desired particle size range.

5.2 BINDERS

The binder is an essential part of a granulation process. The distribution of the binding agent within the granule controls both the intra-granular

How to Optimize Fluid Bed Processing Technology. DOI: http://dx.doi.org/10.1016/B978-0-12-804727-9.00005-3

particulate adhesion, and the inter-granular compression. Most binding agents used for wet granulations are hydrophilic in nature. Binder quality and quantity affect average granule size, granule friability, interparticulate porosity, and granule flowability. The rate of binder addition affects the granule formation and granule size since it affects the degree of wetting and the binder adhesiveness. Increasing the binder addition rate increases the granule size and the granule bulk density due to increased penetration and wetting ability of the binder solution. Reducing the rate of addition will cause the opposite effects, thereby reducing granule size and bulk density of the dried granulation.

Common binders used in the agglomeration process are listed in Table 5.1. Recent addition of a foam binder by DOW Chemical Company provides the option of using polymer foam binder instead of solution of a binder.

5.2.1 How to Select the Proper Binder?

Many different parameters influence the granule properties, but the interactions between the drug and the binder play a very important

Table 5.1 Typical Binders Used in Agglomeration	
Binder	% Used in Formula
Natural polymers	
Starch	2–5
Pre-gelatinized starch	2–5
Gelatin	1–3
Acacia	3–5
Alginic acid	1–5
Sodium alginate	1–3
Synthetic polymers	
Polyvinylpyrrolidone	0.5–5
Methyl cellulose	1–5
HPMC	2–5
Na-CMC	1–5
Ethyl cellulose	1–5
Sugars	
Glucose	2–25
Sucrose	2–25
Sorbitol	2–10

role. A number of techniques can be used to measure wetting and spreading ability for binder solutions to ensure that an appropriate binder is chosen for a particular substrate. Typically, this involves calculation of surface free energies and the works of cohesion, adhesion and spreading (also referred to as spreading coefficient) from measurements of solution contact angles on the substrate of interest, and measurement of the liquid-vapor surface energy of the wetting liquid usually referred to as surface tension.

From a practical standpoint, once the compatibility of the binder and the substrate is established during the pre-formulation screening, the challenge is to determine what percentage of the binder will be needed to form the bond between the particles. Binder efficiency may be defined as the minimum binder use level that is required to achieve a certain benchmark tablet-crushing strength and friability. The robustness of the agglomerated granule will not be determined until the granules are dried. Poor wettability and spreadability of the binder are frequently associated with porous, weak, low-density granules with nonuniform binder distribution and broad particle size distributions. Binders can be added as a powder along with other ingredients in the fluid bed bowl, and the solvent (e.g., water for aqueous granulations) is sprayed to form granules. Depending on the binder, bonds formed by the dry binder are not as strong as if the binder is in solution, because the hydration of polymer takes time. Hence it is advisable to always use a binder in a solution. If you have all of the ingredients or the majority of them are highly water soluble, you may be able to spray only water to granulate the composition in the bowl; in this case, the needed dry binder quantity requirement may be higher.

With a weak or lower amount of binder, the agglomeration will be friable and may not withstand the vigorous fluidization encountered in the fluid bed. On the other hand, if the quantity of the binder in the granules is too high the agglomerated granules will be too hard. This will manifest itself into compression issues such as tablet mottling, or higher hardness and potential for higher disintegration/dissolution of the tablets. In addition to choosing binders with lower surface tensions, one may add surfactants to reduce surface tension or alternately add a less polar organic solvent to water, or completely replace the aqueous solution with a less polar, organic solvent [1]. The concentrations listed in Table 5.1 should be used only as a guideline and assessment of its suitability and its concentration, for the product being granulated should be evaluated as stated above. The binder properties such as viscosity, spreading coefficient solubility, etc., will determine ultimately what quantity of binder and what concentration is needed. The objective should be to make as much concentrated solution as you can make, which will be easily sprayed into atomized droplets to minimize the spraying time, while obtaining desired granules and thus making an optimized process. The impact of the binder properties on the granule morphology was studied by Rajniak et al. [2] qualitatively predicted using simple physically based criteria, which combine the morphological properties of excipients (size and surface roughness) together with physical properties (viscosity, wetting properties, droplet size) of the binder.

5.3 NOZZLES

The degree of atomization of the binder solution is controlled by the proportion of air to liquid mixture in the nozzle head. The mean droplet size is primarily influenced by nozzle construction, air to mass ratio and the dynamic force of atomizing air, by surface tension, density and viscosity of the liquid and density of the atomizing air. Provided that the liquid orifice is wide enough to permit uniform flow, the diameter has no significant influence on droplet size. The nozzle height in relation to the powder bed height in the granulating bowl affects the granule size and granule friability as a result of over-wetting or under-wetting. Considering the extreme conditions, if the nozzle is too close to the fluidized mass, it causes interference in the fluidization pattern. Caking and lump formation will ensue due to over-wetting; worst of all, due to constant impinging of fluidizing particles on the nozzle, it increases the chances of nozzle clogging and disruption of the process. On the contrary, if the nozzle is located at a higher than optimum level, the atomized binder droplets are spray dried before they have a chance to wet the fluidized particles, thereby resulting in under-wetting and therefore reduced agglomeration. The nozzle position is set to give uniform wetting of the "surface" of the bed but with no wall spraying. The droplet size is then set ensuring that they will reach the bed with a minimum of spray drying or blow back.

5.3.1 The Different Orientations of the Spray Nozzle

Top spray: Top-spray granulation has been one of the most recognized and well-studied techniques of fluidized bed granulation since the 1960s. As depicted in Fig. 5.1, the spray nozzle is positioned at the top

Figure 5.1 Top-spray nozzle position. Source: GEA Pharma Systems.

of the product chamber and the binding liquid is sprayed onto the fluidized solid particles, counter-current to the airflow. Granules produced from top-spray granulation are characterized by low bulk density and porous surfaces that promote wicking of liquid into the interstitial voids of the granules, thereby promoting their dispersion and disintegration.

Bottom spray: In this configuration, (Fig. 5.2) the spray nozzle is positioned in the middle of the air distribution plate at the base of the product chamber. A partition column is frequently installed and its presence regulates the fluidization and flow of particles into the spray granulation zone. The binding liquid is sprayed in the same direction as the airflow. This setup is essentially employed for coating purposes and less so for granulation. A modified version of a classical Wurster coater is introduced as "precision coater" (Fig. 5.4) for reducing the process time for coating and also can be used as a granulator.

Rotary and tangential spray: The tangential-spray technique was conceived for producing denser granules than typically possible in fluidized bed granulation. The spray nozzle is introduced at the side of the product chamber and is embedded in the powder bed during processing (Fig. 5.3). The rotating plate of the granulator provides a centrifugal force, which forces the particles toward the wall of the processing chamber.

Figure 5.2 Bottom spray surrounded by partition (cut out). Source: IMA S.p.A.

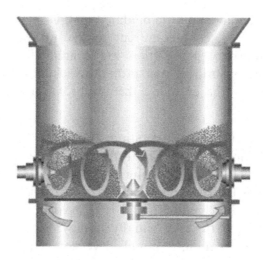

Figure 5.3 Rotary Granulator. Source: Freund-Vector.

The fluidizing air, introduced via a slit, provides a vertical force that lifts the particles upward before gravitational force causes particles to fall down onto the disk. As the granules formed are spherical, denser and less porous than granules produced from the top-spray technique. Rotary processing is suitable for producing granules that are to be coated.

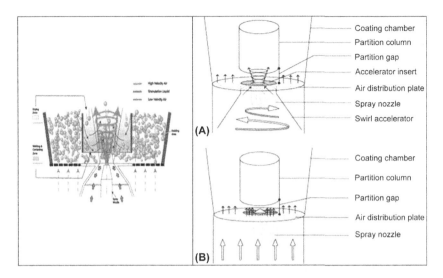

Figure 5.4 Precision Granulator Insert in a fluid bed processor (A) Components of Precision granulator Insert with swirl accelerator. (B) Position of nozzle and the airflow pattern in Precision granulator Insert. Source: GEA Pharma Systems.

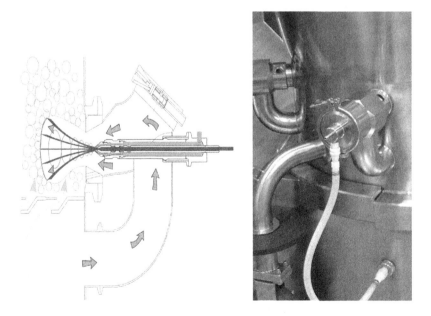

Figure 5.5 Nozzle location in the container of FlexStream system. Source: GEA Pharma Systems.

Modified position of nozzle: Another variation of the nozzle position in the product container is offered by GEA Pharma Systems called "Flexstream" (Fig. 5.5). It offers one product container for all unit operations.

Table 5.2 Advantages and Disadvantages of Various Fluid Bed Granulation Options		
Fluid Bed Granulation Module	**Advantages**	**Disadvantages**
Top spray	• More porous granules • Granules are irregular in shape • "One-pot" system for contained production	• More spray drying of binder solution
Bottom spray	• Smaller granules • Spherical granules with better flow • Smoother surface granules • Less spray drying of binder solution • Robust process at low spray rate	• Need reasonably well flowing feed powders • Denser granules • Slower granule growth • Higher risk of overwetting
Tangential spray	• Supports higher spray rate • Less dependent on flow properties of powder feed Powder elutriation only at higher airflow • Granules have good flow properties • Granules have low friability • Homogeneous drug distribution	• Denser granules • Material loss due to adhesion to friction plate • Scale-up designs are expensive and impractical • Granule growth is relatively uniform • Risk of bed overwetting is high

Advantages and disadvantages of various fluid bed granulation options are listed in Table 5.2.

Solution delivery system: The type of solution delivery system employed will depend on the granulating solution being used. Higher viscosity solutions with solids may work better with a positive displacement pump. Solutions with a large quantity of solids may work better with a weigh feedback system. However, most granulating solutions will work with a mass flow rate system.

5.3.2 Nozzle Blockage

It is also very important for the nozzle to operate without nozzle blockage. Blocking the nozzle during operation leads to deterioration of atomization and possibly to wet or dry quenching. At start-up, bed particles may be blown into the nozzle or the liquid feed may evaporate in a hot nozzle. Blockage may also occur due to evaporation within the nozzle during a run. The ability to purge atomizing air ports could prevent nozzle blockage. Additionally, at the end of the run, flushing the nozzle with a pure solvent will assure the nozzle remains clear of any polymer or product residue.

The solution to be sprayed must be lump-free so that lumps do not block the flow of liquid. Lumps or un-hydrated polymer (fisheyes) account for a large portion of all nozzle clogs in fluid-bed granulators.

It is also possible that as lumps move out of the solution tank and toward the nozzle, they may block one or more ports. Lumps that have migrated to the nozzle port usually attach internally and do not pass into the product bowl. Hence, it is imperative that these clogs be detected and removed as soon as possible.

In a multiport nozzle, an initial clog in a single port often combines with powder and migrates to a second (and perhaps still another) port. In other cases, after the solution hardens at the nozzle surface, it migrates downward and forms a "beard" or "ice pick." Once the beard is formed, the solution pattern is deflected downward immediately below the nozzle. This is often evidenced by a conical-shaped adhesion of solution on the product bowl screen. Damaged "O" rings are also frequently responsible for improper needle seating, which can produce the effects described above. It is equally important to check the needle valve assembly regularly and to replace damaged or worn O rings as required. Frequent bag shaking during the spray cycle triggers the compressed-air purge mechanism that also provides frequent purges of solution jets. This is an effective measure to prevent nozzle clogging. Nozzle atomizing pressure actually assumes the nozzle is also receiving the required mass flow rate of air. If the pressure is monitored close to the nozzle, then the mass flow rate of atomizing air is most likely correct. However, this would miss the fact that the nozzle is not clean at the tip and the nozzle is not operating properly.

5.4 PROCESS PARAMETERS

The fluidized bed process has a multitude of variables, most of which are confounding, and several of which will have a strong impact on dosage form performance. Table 5.3 lists typical process parameters for the fluid bed granulation and Table 5.4 shows the impact of key operating variables during the fluid bed granulation.

Table 5.3 Typical Process Parameters in Fluid Bed Granulation		
Process Air	**Spray**	**Other**
Incoming air dew point	Spray rate	Bed depth
Process air sew point	Atomization air pressure for nozzle	Batch size
Inlet air temperature	Atomizing air volume	Bowl screen area
Total process air volume	Viscosity of solution	Process filter type and porosity
Process air velocity	Nozzle height	Filter shake interval and time
	Number of nozzles	Product pressure drop
	Nozzle set up (port size/air cap)	Process filter pressure drop

Table 5.4 Impact of Key Operating Variables

Impact of Key Operating Variables in Fluid Bed Granulation

Operating or Material Variables	Effect of Increasing Variable
Increasing solids mixing, solids flux, and bed agitation	Increasing gas velocity Improves bed uniformity Increases solids flux Decreases solid circulation time Potentially improves nucleation Increases growth rate Lowers growth limit Increases granule consolidation Increases granule attrition Increases initial drying kinetics Distributor design Impacts attrition and defluidization
Increasing bed weight (and height)	Increases granule consolidation, density, and strength
Increasing bed moisture Increasing residence time	Increases rates of nucleation, growth and consolidation- larger denser with wider PSD Distribution can narrow if growth limit is reached Increases chances of defluidization
Improving spray distribution Lower liquid feed or spray rate Lower drop size	Largely affected: Wettable powders and shorter penetration times required For fast penetration Decreases growth rate Decreases spread of size distribution Decreases granule density and strength
Increasing feed particle size	Requires increase in excess gas velocity Minimal effect of growth rate Increase in granule consolidation and density
Spray rate or increase in number of nozzles	Increase size and spread of granule size distribution Increase granule density and strength Increase chance of defluidization due to quenching
Liquid droplet size	Increase size and spread of granule size distribution
Air velocity (increase in airflow)	Increase attrition and elutriation rates Decrease coalescence for inertial growth No effect on coalescence for noninertial growth Increase granule consolidation and density
Bed temperature	Decrease granule density and strength Finer granules
Binder viscosity	Increase coalescence for inertial growth No effect on coalescence for noninertial growth Decrease granule density

Source: *Adopted from [3].*

5.5 CHALLENGES IN FLUID BED GRANULATION

Table 5.5 lists some of the common challenges during the fluid bed granulation and probable causes and possible solutions to overcome them. *(Also see Chapter 13.)*

Table 5.5 Challenges During Fluid Bed Granulation and Probable Cause and Possible Solution

No.	Observation/Challenge	Probable Cause and Solution
1	*Excessively coarse granules*	• Inlet air temperature too low • High spray rate, pump calibration • Nozzle position too low • Atomization air is not on and binder does not atomize • Nozzle leakage • Surface area and the water absorption properties of the starting material essentially influence the granule growth.
2	*Excessive fines*	• Inlet air temperature is too high • Binder spray rate is too low • Insufficient quantity of binder • High fluidization velocity or airflow • Nozzle position too high
3	*Dried product has combination of coarse and fine granules*	• A poorly functioning spray nozzle will typically cause a combination of fines and coarse, dense granules. It does so as a consequence of nonuniformity of droplets—the majority is a fine mist, but there is likely to be a component of very large droplets (exceeding 50 microns) that form granules with nearly liquid centers. The resulting particle size distribution may be bimodal. The consequent dense granules will result in nonuniformity of moisture distribution because they possess little interstitial porosity. • Nozzle height influences the granule size distribution
4	*Final moisture inconsistency*	• Inadequate process development drying curve • Improper fluidization • Temperature probe out of calibration
5	*Poor fluidization*	• Too much product in the product container • Incorrect air distributor plate • Processor fan does not have adequate pressure drop • Air distributor not cleaned properly • Exhaust filter porosity too small • Exhaust filter is blocked
6	*Low yield*	• Wrong porosity exhaust filter • Air distributor with coarser screen opening • Filter bag with a tear in it • Filter bag not shaken at the end of the process • Material sticks to the walls of the expansion chamber as a result of static charge.
7	*Finished product nonuniformity*	• Insufficient filter shaking • Product homogeneity before granulation not adequate • Lumpy raw materials • Spraying time insufficient
8	*Difficulty in fluidizing very fine and low-density products due to static electricity*	• Buildup of static electricity on the particles is quite a common problem especially when a fine powder is fluidized. This can lead to uneven fluidization and/or sticking on the bed walls, spray nozzles, etc. Static problems can be overcome by increasing the moisture content of the fluidized air stream. • Minimize fluidization time before starting the spray. • Also, as the spraying starts, the static charges will be minimized as the humidity in the processor increases.

(Continued)

Table 5.5 (Continued)

No.	Observation/Challenge	Probable Cause and Solution
9	*Granules are "case hardened" (dried from outside and wet inside)*	• If larger and relatively harder granules are formed and the composition produces dried outer surface with wet center
	Wet gets too wet and spray rate cannot be changed to optimize the process.	• Reduce the spray rate or increase the inlet air temperature and reduce the incoming air humidity (lower the dew point). • There are both mass and energy balance limits on the maximum liquid feed rate. • Exhaust air humidity cannot exceed the saturation humidity at the exhaust temperature. Once the exhaust air is saturated no more liquid can be removed from the bed. • Energy required to evaporate the liquid cannot exceed that available from the inlet air temperature and in turn capacity of the air heater.
10	*Segregation of particles in FB*	• The extent of fluidization segregation is dependent upon a combination of material properties, process equipment characteristics, and process conditions. • Segregation risk is minimized when the physical properties of a mixture's individual components are most similar. • Cohesion also plays an important role where sticking together of individual and dissimilar particles can minimize segregation by limiting de-mixing.
11	*Granulating in a fluid bed does not produce granules at the end of process*	• High temperature of incoming air • Low spray rate • Wrong binder used • Quantity of binder is insufficient • Type and concentration of binder affects granule size.
12	*Granulating solvent is not water but organic solvent*	• Make sure all process equipment is explosion proof, well grounded and the room has proper classification. • System is equipped with either catalytic oxidizer or scrubber to process effluent organic vapors.
13	*Product is highly potent compound or controlled drug substance*	• Assure the operator exposure is eliminated by using personal protection as well as isolator if necessary. • Make sure fluid bed processor has filters (preferably stainless steel) and the processor can be cleaned by Clean In Place system.
14	*API quantity is very small and content uniformity is not acceptable*	• Consider dissolving API in the solvent and spray
15	*For a multiport nozzle, total spray volume for each nozzle cannot be calculated*	• Use one pump for one nozzle with a flow meter • Multiple nozzles will require multiple pumps with manifold
16	*Atomization air has oil in the line*	• Make sure the compressor supplying atomization air is oil free • Install oil and moisture filters in line and periodically clean them
17	*Nozzle spraying on the wall*	• Make sure the nozzle angle is correct and covers the bed
18	*Nozzle clogging during processing*	• Follow nozzle cleaning procedure • The fine powder has a tendency to enter into the needle air venting hole located on the closure cap. The fine powder, once it enters the cap, mixes with the spray medium and cakes or solidifies. • This prevents the needle plunger from operating correctly. • The cap should be removed and the cavity checked for this possibility after every run.

(Continued)

No.	Observation/Challenge	Probable Cause and Solution
	Table 5.5 (Continued)	
19	*Nozzle is leaking has erratic pulsation*	• Maybe one of the "O" rings is damaged. • A properly functioning nozzle should have a spray pattern that is free from erratic pulsation. If a peristaltic pump is used, there will likely be a rhythmic pulse as the lobes of the pump engage the tubing. This is characteristic of this type of pump and not a defect. • Every time a nozzle has to be used, spray test should be conducted at the spray rate and atomizing air pressure that will be used for the slowest spray rate in the process. Any leaks or pulsation will be evident before the nozzle is placed in the processor.
20	*Pump tubing is jiggling erratically*	• This is a sign that compressed air is "leaking" into the liquid. This will cause a significant distortion in the spray pattern, and agglomeration (in particle coating or layering) is a virtual certainty.
21	*Process filters clogged up*	• Wrong micron opening of the filter porosity-check the particle size of the product being processed and choose appropriate micron rating for the process filters • Too high airflow • Product has low density • Change the air distributor to control the air velocity.

5.6 GRANULATION END POINT

5.6.1 Off-Line Methods

Fluid bed granulation is a dynamic environment and measuring the end point of granulation is difficult. However, the true end point and the particle size determination are evident after the drying is complete. So the granule formation during the binder spray may or may not remain, as granules, will depend on the amount and the mechanical strength of the binder along with solubility of the ingredients as they form the bonds during the spraying stage. In determining the end point, granule porosity, flow properties, density, and the particle size distribution are the critical properties should be the criteria for completion of the process. The end point of the granulation process is determined based on the amount of binder liquid added as the primary particles agglomerate. Normally during the product development stage, the quality of final granule characteristics is determined. The scale-up process needs to be robust enough to be able to scale up the quantity of binder liquid as well. The quality of the manufactured granules is normally assessed via (time-consuming) laboratory analysis of the critical quality attributes of selected samples. A more efficient way to control the granulation process consists of the real-time product quality assessment, supplemented with real-time process parameter adjustments correcting for undesired changes in the product properties and process progress. This entails the development of automated granulation processes that use in-line devices to directly measure the critical product parameters.

5.6.2 Near Infrared (NIR)

The moisture content and particle size determined by the near-infrared monitor correlates well with off-line moisture content and particle size measurements. Given a known formulation, with predefined parameters for peak moisture content, final moisture content, and final granule size, the near-infrared monitoring system can be used to control a fluidized bed granulation by determining when the binder addition should be stopped and when the drying of the granules is complete [4]. Near Infrared (NIR) spectroscopy has been shown to be promising in measuring particle size due to its cross sensitivity. Particle size information can be extracted from baseline offset and slope change in the NIR spectra through chemometric modeling [4–9]. Several studies showed that combining the NIR moisture information with the traditionally collected process parameters increased granulation information, improved the predictability of models and contributed to the process optimization [10,11]. The challenge with this online technique is that the measurement device is at the fixed location while the particles are flying all over the process chamber. Few techniques are commercially available for in-line particle size measurement during wet granulation [12,13]. NIR measurement with a diode ray detector has been introduced recently. It avoids window fouling. The Diode Array Detectors in the unit can take spectra in a few milliseconds, by averaging these spectra over time, to adjust the measured volume of product to what is normally used in off-line measurements. The Online LOD sensor can measure the moisture content of within a range of 1%–20% with a projected accuracy of ± 0.25% subject to calibration verification and suitable compendial method [14].

5.6.3 Focused Beam Reflectance Measurement (FBRM)

The focused beam reflectance measurement (FBRM) instrument is designed to track in real-time any changes to the particle size and its distribution. The FBRM probe scans with a focused beam of laser light in a circular path at a high speed (2–8 m/s). Particles passing in front of the measurement window are hit by the laser light, which causes the scattering of the laser light in all directions. The light backscattered into the probe is used to calculate particle chord length and particle chord length distribution. The at-line FBRM application shows the granule growth kinetics, but the fouling that may be observed during in-process measurements impedes the reliability of the FBRM technique as an in-line process analyzer. Adaptations to the probe have been made to solve these fouling issues. The FBRM C35 probe is equipped with a pressurized air-activated mechanical scraper on the sapphire measurement window to prevent powder sticking.

5.6.4 Spatial Filtering Velocimetry (Parsum)

Besides FBRM, spatial filtering velocimetry (Parsum) can be used for measuring the particle size. *(See Chapter 10.)* The Parsum probe uses the measurement principle of spatial filtering velocimetry. This is a number-based, chord length sizing method that collects data for individual particles to obtain a particle size distribution and velocity distribution. Particles falling through the Parsum probe measurement zone block the light emitted from a laser light source. The resulting shadow is detected using a linear detector array made up of optical fibers. Jun Huang et al. used Parsum probe to evaluate in-line monitoring manufacturability quality attributes, particle size, and particle size distribution [15].

5.6.5 Imaging Methods

Images give direct information on particles. The image analysis process is composed of five steps: image acquisition, preprocessing, segmentation, extraction, and representation of the characteristic parameters. Watano and Miyanami [16] used image analyses in a real-time particle size determination of agitation fluid bed granulation. Recently, Eycon has introduced an image unit that captures images and subsequently calculates particle size distributions for the sample materials in a rapid and accurate manner.

5.7 GRANULATION CHARACTERIZATION

- *Friability*: Abrasion resistance or friability has been used to evaluate granule strength. A friability index (FI) has been numerically calculated by using a friability test as the change in the mean particle size before and after the test. The FI is a parameter that defines a single point. However, the friability event of the granules shows continuity as a function of time during the sieving and mixing periods after the drying process at the industrial manufacturing stage. There is no official procedure for the testing of granule friability.

Table 5.6 Off-Line Procedures to Characterize the Granules		
No	Parameters	Method
1	Particle morphology	Optical microscopy
2	Particle size distribution	Sieve analysis, laser light scattering
3	Powder nature	X-ray diffraction
4	Thermal analysis	DSC, TGA, DTA5
5	Identification	NIR spectroscopy
6	Surface area	Gas absorption
7	Granule porosity	Mercury intrusion methods
8	Followability of granule	Angle of repose
9	Density	Density apparatus

- *True density*: True density was measured using a helium pycnometer. Results are averages of five replicate determinations.
- *Bulk density*: A sample is gently passed into a 100 mL graduated cylinder to the 100 mL mark and weighed. From the mass and volume data, the bulk density is calculated. Results are averages of three replicate determinations.
- *Tapped bulk density*: The same sample used for bulk density measurement can be subjected to a tapped density tester. A total of 1500 taps with the displacement amplitude of 14 mm is used for the determination. Results are averages of three replicate determinations.
- *Flow property determination*: The flow rate is determined with a flow tester by weighing the same quantity at each time of friable and un-friable granules. The angles of repose of friable and un-friable granules were determined by the dynamic angle of repose method by using 100 mL of the granules.

Some of the off-line procedures used to characterize the granulation are summarized in Table 5.6.

5.8 SUMMARY

Fluid bed granulation is a critical unit operation. Understanding the equipment and functionality and limitations of each component is essential to develop the fluid bed granulation process. The interactions of the process variables create a dynamic environment where the quality of granules produced depends on the fluid bed processing parameters as well as the nozzle operation and the binder used.

Understanding the complexity of this unit operation will help develop and optimize the granulation process.

REFERENCES

[1] Durig T. Binders in pharmaceutical granulation. In: Parikh DM, editor. Handbook of pharmaceutical granulation technology. New York: Informa Health; 2009.

[2] Rajniak P, Mancinelli C, Chern RT, Stepanek F, Farber L, Hill BT. Experimental study of wet granulation in fluidized bed: impact of the binder properties on the granule morphology. Int J Pharm 2007;334:92−102.

[3] Ennis B. Theory of granulation and engineering perspective. In: Parikh DM, editor. Handbook of pharmaceutical granulation technology. 3rd ed. New York: Informa Healthcare; 2009. p. 54.

[4] Findlay WP, Peck GR, Morris KR. Determination of fluidized bed granulation end point using near-infrared spectroscopy and phenomenological analysis. J Pharm Sci 2005;94:604−12.

[5] Rantanen J, Wikström H, Turner R, Taylor LS. Use of in-line near-infrared spectroscopy in combination with chemometrics for improved understanding of pharmaceutical processes. Anal Chem 2005;77:556−63.

[6] Frake P, Greenhalgh D, Grierson SM, Hempenstall JM, Rudd DR. Process control and end-point determination of a fluid bed granulation by application of near infra-red spectroscopy. Int J Pharm 1997;151:75−80.

[7] Goebel SG, Steffens KJ. Online-measurement of moisture and particle size in the fluidized-bed processing with the near-infrared spectroscopy. Pharm Ind 1998;60:889−95.

[8] Nieuwmeyer FJS, Damen M, Gerich A, Rusmini F, Maarchalk KV, Vromans H. Granule characterization during fluid bed drying by development of a near-infrared method to determine water content and median granule size. Pharm Res 2007;24:1854−61.

[9] Rantanen J, Räsänen E, Tenhunen J, Känsäkoski M, Mannermaa JP, Yliruusi J. In-line moisture measurement during granulation with a four-wavelength near infrared sensor: an evaluation of particle size and binder effects. Eur J Pharm Biopharm 2000;50:271−6.

[10] Parikh DM, Bonck JA, Mogavero M. Batch fluid bed granulation. In: Parikh DM, editor. Handbook of pharmaceutical granulation technology. New York: Marcel Dekker Inc.; 1997. p. 227−302.

[11] Parikh DM. Batch size increase in fluid bed granulation. In: Levin M, editor. Pharmaceutical process scale-up. New York: Marcel Dekker Inc.; 2002. p. 171−220.

[12] Näivänen T, Lipsanen T, Antikainen O, Räikkönen H, Heinämäki J, Yliruusi J. Gaining fluid bed process understanding by in-line particle size analysis. J Pharm Sci 2009;98:1110−17.

[13] Schmidt-Lehr S, Moritz H, Jürgens KC. Online control of particle size during fluidized bed granulation. Pharm Ind 2007;69:478−841 2.

[14] "Lighthouse probe" by GEA, <http://www.gea.com/global/en/products/lighthouseprobe.jsp>.

[15] Huang, et al. A PAT approach to enhance process understanding of fluid bed granulation using in-line particle size characterization and multivariate analysis. J Pharm Innov 2010;5:58−68. Available from: http://dx.doi.org/10.1007/s12247-010-9079-x.

[16] Watano S, Miyanami K. Image processing for on-line monitoring of granule size distribution and shape in fluidized bed granulation. Powder Technol 1995;83:55−60.

CHAPTER 6

Drying

6.1 THEORY

The removal of moisture from a product granulated in the fluid bed granulator or in other equipment essentially removes the added water or solvent. This free moisture content is the amount of moisture that can be removed from the material by drying at a specified temperature and humidity.

During fluid bed granulation, spraying of liquid and evaporation of solvent are taking place simultaneously. When binder addition is stopped, the drying of granules continues until the unbound moisture is evaporated. The end of drying is indicated when the surface moisture is replaced by diffusion, from the center of the granules through capillary, resulting in the rise of the product temperature. For product granulated in a mixer, the sequence of drying takes place similar to the product drying stage in fluid bed.

The fluidized bed drying process is a three-stage process, including a short preheating stage, a constant rate stage, and a falling rate stage. The constant rate stage corresponds to a constant bed temperature. The material is dried while suspended in the upward-moving conditioned air. The capacity of the air (gas) to absorb and carry away moisture determines the drying rate and establishes the duration of the drying cycle. Unbound water is easily lost by evaporation until the equilibrium moisture content (EMC) of the solid is reached. Once the solid reaches its EMC, extending the time of drying will not change the moisture content as an equilibrium situation has been reached. The higher the temperature of the drying air, the greater its vapor holding capacity. However, the maximum drying air temperature will be determined by the heat sensitivity and thermoplasticity of the product. As in heat-transfer, the maximum rate of mass transfer that occurs during drying is proportional to the surface area, turbulence of the drying air, the driving force between the solid, and the air and the drying rate. Since, the temperature of the wet granules in a hot air depends on the rate of evaporation, the key to analyzing the drying process is psychrometry.

How to Optimize Fluid Bed Processing Technology. DOI: http://dx.doi.org/10.1016/B978-0-12-804727-9.00006-5

The fluid bed dryer provides significant advantages over the conventional tray dryer traditionally used in the pharmaceutical industry. These advantages include:

1. increased rates of drying and product throughput, accompanied by significant improvements in thermal efficiency
2. increased drying capacities per unit of floor space
3. increased ability to control product temperature during drying and the facilitation of the handling of heat sensitive materials
4. decreased handling costs resulting from simplified loading and unloading operations.

6.2 DRYING GRANULATED PRODUCT

Product granulated in either high shear or low shear mixers can be dried in a tray dryer or fluid bed dryer. For process efficiency and optimization, fluid bed drying is preferred for the reasons mentioned above. During drying, the inlet air temperature and humidity, product and exhaust air temperature, and airflow are continuously monitored. In particular, the product and exhaust air temperature indicate the constant rate and falling rate of drying periods. Fig. 6.1 shows a typical drying curve of fluid bed drying.

6.2.1 Drying Challenges
- Loading the wet granulated product in the fluid bed bowl
- Effect of particle size on drying end point
- Minimizing granule breakage
- Fluidizing heavy/cohesive product

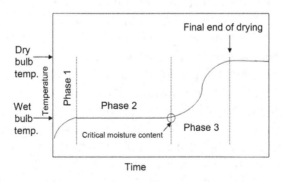

Figure 6.1 Temperature phases in the drying process. From Parikh DM. Batch fluid bed granulation. Chapter 10, Handbook of pharmaceutical granulation technology. 3rd ed. New York: Informa healthcare; 221: 2009 [1].

- Minimizing product segregation
- Avoiding uneven drying of the granulation
- Drying nonaqueous granulation
- Minimizing the drying time to increase optimizing the drying process
- Increasing yield after drying
- Determination of end point

6.2.1.1 Loading the Wet Granulated Product in the Fluid Bed Bowl

Charging of the granulated product in the fluid bed bowl can be accomplished either by manually discharging the product from the mixer in a fluid bed bowl or pneumatically transporting the product directly to the fluid bed, as is generally done in an integrated system setup (see Chapter 12).

6.2.1.2 Effect of Particle Size on Drying End Point

For drying granulated product, it is advisable to pass the wet mass through a coarse screen, so the cohesive chunks of product can be subdivided into free flowing granules. This will result in larger surface area available for drying resulting in faster drying, improving the process efficiency. This can also minimize the "case hardening" of granules where the outer surface dries but the center of the granule stays moist, creating further processing problems.

6.2.1.3 Minimizing Granule Breakage

In fluid bed processing, intense mixing and particle-to-particle collision, particle-to-vessel wall collision takes place. Breakage and attrition of granules during drying reduces the particle size as the product dries, which generates fines and affects the flow properties of dried product. Formation of fines by attrition or abrasion is, in practice, an important parameter because it can affect flowability of the granule mass. This breakage depends upon the strength of the formed granule, which in turn reflects the amount and kind of binder used during granulation. Normally reducing the fluidization velocity at the drying stage will help reducing the generation of fines. Granules can exhibit an intrinsic breakage propensity during drying depending on the water content, type of binder and extent of stress exposed to the granules.

6.2.1.4 Fluidizing Heavy/Cohesive Product

When drying cohesive materials in the fluid bed, the inter-particle forces are considerable and they control the behavior of a bed. During the fluidization, the bed cracks into large portions and the gas tends to flow into the gap between the fissures. Then, channeling occurs in the bed, and eventually, the gas solid contact is very low and heat and mass transfer operation is weakened. Cohesive granulation or product from centrifuge (in case of API) sometimes is difficult to fluidize with just process air. It may be necessary to provide a mechanical rake in the product bowl to impart mechanical energy and break up the cohesive product until fluidization air is capable of fluidizing the product. In some cases, distributor plate replacement may be necessary. A restrictive distributor will provide higher fluidization velocity at a given air volume to facilitate the fluidization of the product. Another option is to pass the cohesive product through the mill before charging into the fluid bed processor. An integrated system as discussed in Chapter 12 may help understand how to transfer product from a mixer via mill to the running fluid bed.

6.2.1.5 Minimizing Product Segregation
Product segregation of already granulated product during drying will depend on the physical properties of the ingredients like solubility, particle size, density, flowability, solvent type, and strength and quantity of the binder. The property of the major component will have an impact on whether segregation will take place. The combination of granulating solvent and drying conditions could result in conversion of some of the products to alternate crystalline forms during the drying process.

6.2.1.6 Avoiding Uneven Drying of the Granulation
Inefficient or poor control of the drying process will lead to products of variable quality. This might occur if there is a variation in the particle sizes, binder solvent, density, and particle size of the granules being dried. This may also result due to the uneven fluidization velocity caused by a blocked distributor from fines or lack of proper cleaning. A wrong distributor may be another reason for uneven fluidization. Excessive velocity indicated by the rapid rise in the exhaust air temperature could result in nonuniform drying of the product besides resulting in an inefficient process.

6.2.1.7 Drying Nonaqueous Granulation
In a case of granulation requiring flammable solvents, process air and nozzle atomization air are replaced by an inert gas such as nitrogen and the system is designed as a closed cycle with the solvent recovery

capability. Alternatively, when drying solvent granulation, it is a normal practice to modulate volume of inlet air to keep solvent vapors in the exhaust air stream remain below lower explosion limit (LEL) and use of thermal oxidizer to burn solvent vapors in the exhaust air. Generally organic solvents, due to their rapid vaporization from the process, produce smaller granules than the aqueous solution. Caution must be taken to utilize sufficient airflow to overcome the solvent's lower explosion limit. Aqueous drying is very different from organic solvent drying. Low vapor pressure, high heat of vaporization, and the impact of inlet air humidity combine to make aqueous drying more difficult than nonaqueous product drying.

6.2.1.8 Minimizing the Drying Time to Increase Optimizing the Drying Process

The efficiency of drying can be improved and the process time reduced by increasing the inlet air temperature, because the high temperature has a higher capacity to hold water. For larger batch sizes, higher inlet air temperature is beneficial to counter higher weight of the batch. The higher weight of the batch may cause compaction of the porous granules. Higher temperature will produce more porous granules and shorter drying time. The use of higher inlet air temperatures was reported to enhance evaporation rate; the evaporation rate increased approximately from 56 to 78 g/s when inlet air temperature was raised from 90 to 120°C at a fixed airflow rate of 4000 scfm [2]. An increase in the particle diameter decreases the available particle surface area and results in lowering the drying rate for coarse sized particles compared with fine sizes; hence, milling the granulated product through a coarse mill prior to drying is highly recommended. The humidity of the inlet air influences the drying rate of aqueous-based granulations [3]. Studies have shown that an increase in inlet air humidity results in higher product temperatures [4,5]. A linear temperature increase of 10°C was observed in the powder bed when the absolute inlet air humidity was increased from 4.5 to 23.4 g/m^3 during the liquid addition phase [4].

6.2.1.9 Increasing Yield After Drying

Depending on the final product particle size, which could be the result of attrition, friability of granule, and filter porosity, product yield from drying could be reduced by 3%−5%. Besides robust formulation at the granulation stage, for reducing the breakage of granules, several options should be considered such as: lowering the fluidization air volume toward the end of the drying cycle, selecting the appropriate porosity of the process filters, and selecting the appropriate air distributors. In the case of a product manufactured in a multiple batch campaign, the product yield from the first batch may be lower than the batches manufactured subsequently.

6.2.1.10 Determination of End Point

Most pharmaceutical solids have poor thermal conductivity; thereby slow terminal diffusional rates add drying costs when very small amounts of moisture in the end product are a target. Generally, a desired moisture level of the dried granulation determines the end point of the drying process. There are many methods for monitoring the fluidized bed drying end point.

6.2.1.10.1 Off-Line Methods

• Temperature Monitoring:

The most common practice is to monitor the exhaust air temperature to indicate the unbound moisture removal and increase in exhaust temperature. The method involves removing samples from the bed at the beginning of drying and subsequently at a predetermined time interval, and testing the sample for loss on drying (LOD) using a moisture balance. By noting the temperature of exhaust at each time the sample is taken, a table can be prepared which corresponds to the LOD reading. The procedure is carried out until the product is dried to the required moisture content. The exhaust temperature at this point can be used as the end point. To assure the robustness of this approach, several batches will have to be monitored and moisture level checked against the exhaust air temperature. A similar procedure could be used to monitor and record the product temperatures and establish a similar drying curve.

• Humidity Measurement

Another temperature-based method for monitoring drying involves estimating the humidity of the air leaving the dryer by comparing the wet bulb and the dry bulb temperatures. Initially, the humidity of the air leaving the dryer is very high due to evaporation of water from the granules during drying. As the granules dry, the humidity of the outlet air decreases and the end point of drying occurs when the humidity of the inlet and outlet air approach the same value.

6.2.1.10.2 Other Off-Line Methods

Off-line analyses (LOD in an oven, Karl Fisher titrations, and gas chromatography) are more accurate but significantly increase cycle time. The off-line methods have disadvantages such as limited accuracy due to poor fluidization conditions within the bed. The requirement of manual sampling of wet products during the drying phase may lead to potential safety and industrial hygiene issues such as operator exposure to potent drugs. If the attainment of a specific product temperature was used as the end point of drying, and granule particle size is fairly high, then it is possible that the dried granules are case hardened and may contain an undesirably high amount of residual moisture that may adversely affect product quality. These findings are particularly important in the drying of moisture-sensitive materials.

6.2.1.10.3 On-Line/At-Line Methods

Near-Infrared (NIR)

Near-infrared (NIR) spectroscopy has been used to monitor drying by measuring the moisture content of the air moving through the dryer and, more commonly, by measuring the moisture content of the granules within the bed. Quantifying water in most compounds is very easy and straightforward with NIR. This is due to the fact that water is the most sensitive compound that can be measured by NIR. Most compounds can be detected reliably down to 0.05%–0.1% while water sensitivity is roughly an order of magnitude better. NIR spectra are characterized by wide and overlapping peaks which are often visually difficult to interpret as they are not bond specific. Thus, the determination of chemical and physical properties by NIR involves the use of multivariate calibration tools to model the property of interest. The importance of NIR spectroscopy is increasing as the tool used for monitoring the drying process of granules. Calibration models have been developed by comparing the NIR algorithm with off-line analytical measurements of moisture levels and particle size. Since granular particle size decreases during the drying process due to collisional forces, it is important to measure both the moisture level and granule size. Apart from determining the end point for drying of granulation using moisture during the process, NIR can also be used for drug content and hardness determination of the dried granules at the end of drying stage. Table 6.1 shows the comparison between on-line and off-line methods for moisture measurement. (*Chapter 10 provides application of NIR for determination of Moisture.*)

Table 6.1 Advantages of FT-NIR Over Laboratory Moisture Analysis	
FT-NIR	**Karl Fischer/Loss on Drying**
Nondestructive	Destructive
Uninterrupted sampling	Interrupted sampling
No operator training needed for analysis	Chemistry and method training for operator
No sample preparation or solvents	Sample preparation and solvents
Results in seconds	Results in 15 min or more
Allows for closed loop control	No ability for closed loop control
Multi-component analysis on dryer samples	Additional tests on samples would take a long time
NIR can be used online, inline or at-line	Can only be run in the laboratory—can't go online
Source: *Thermo Fisher Antaris™ MX FT-NIR process analyzer brochure, [6].*	

Effusivity

Thermal effusivity is a material property that combines thermal conductivity, density, and heat capacity. Hence, it can differentiate between solids, liquids, and powder components in a system based on heat-transfer properties. Water has the highest effusivity value (1600 Ws½ (m²k)) and air has the lowest value (5.5 Ws½ (m²k)). Effusivity sensors identify an ingredient on the basis of that material's unique properties. Initially, the effusivity equipment can be used off-line until full scale trials are warranted. The off-line measurement requires that the samples are extracted from the process and brought to the effusivity instrument to be tested.

The on-line technique uses an automated system that extracts the sample from the dryer. The material needs to be in contact with the effusivity sensor for a short period of time (1–10 s), which subsequently tests it under fixed pressure, and then returns it to the dryer if required. This requires modification of the sample port for the sensors to be placed in the fluid bed unit. Since the effusivity of water is much higher than the pharmaceutical solids (150–800 Ws½ (m²k)), effusivity measurement during drying is particularly sensitive to moisture changes and thus can be used to measure end of drying. Off-line and at-line measurement of effusivity can be correlated to the moisture content of the product [7].

6.3 SUMMARY

Fluid bed drying is a critical unit operation for producing pharmaceutical granulations. Understanding the criticality of multivariate nature process parameters along with the capability to configure the equipment needed for the process will provide optimized drying results. There are different challenges when the product is granulated in a high or low shear mixer and fluid bed dried; these challenges can be overcome with the understanding of the confounding nature of fluid bed parameters. The end point of drying is the most important attribute required because it has implications for subsequent processing and stability of the granulation. Various off-line and on-line methods are available and need to be implemented to achieve the optimized drying process.

REFERENCES

[1] Parikh DM. Batch fluid bed granulation. Chapter 10, Handbook of pharmaceutical granulation technology. 3rd ed. New York: Informa healthcare; 2009.

[2] Hlinak AJ, Saleki-Gerhardt A. An evaluation of fluid bed drying of aqueous granulations. Pharm Dev Technol 2000;5(1):11−17.

[3] Zoglio MA, Streng WH, Carstensen JT. Diffusion-model for fluidized-bed drying. J Pharm Sci 1975;64(11):1869−73.

[4] Lipsanen T, Antikainen O, Raikkonen H, et al. Novel description of a design space for fluidised bed Granulation. Int J Pharm 2007;345(1−2):101−7.

[5] Lipsanen T, Antikainen O, Raikkonen H, et al. Effect of fluidization activity on end-point detection of a fluid bed drying process. Int J Pharm 2008;357(1−2):37−43.

[6] Thermo Fisher Antaris™ MX FT-NIR process analyzer brochure.

[7] Keintz R, Fariss G, Okoye P. Thermal effusivity and power consumption as PAT tools for monitoring granulation end point. Pharm Technol 2006;30(6).

CHAPTER 7

Coating

7.1 INTRODUCTION

Coating of particles or pellets is required for various reasons such as taste masking, moisture protection or for modified release dosage forms. Fluid bed coating offers the possibility to alter and to improve various characteristics of core particles such as the surface properties in a single unit operation. The quality of the coating extensively depends on the statistical residence time of the particles in the coating zone. Coating these pellets could be accomplished by placing the nozzle in a fluid bed unit from the top (top-spray), from the bottom (Wurster technique), tangentially as in a rotary fluid bed module or from the side as offered by the FlexStream module (GEA Pharma Systems.).

7.2 TOP-SPRAY COATING

Coating with the top position of the nozzle is like the top-spray granulation. In a conventional fluid bed granulator, particles circulate within the bed but in a less ordered or more chaotic manner than in the conventional Wurster technique with a column. The coating liquid is nearly always sprayed downwards from the top of the bed, which has the advantage of the nozzle being removable during operation if it becomes clogged. Due to the less ordered circulation of particles, this technique is rarely used for sustained or enteric coatings, although application of moisture barriers and cosmetic coats are common. This technique has also been used successfully to apply hot melt (solventless) coatings. The top-spray system is most effective when coatings are applied from aqueous solutions, latexes, or hot melts. This process is generally used for taste masking, enteric coating or applying moisture barrier film. However, the top-spray is not suitable for sustained release dosage forms when precise reproducibility is required or where organic solvent-based enteric coating is to be used.

How to Optimize Fluid Bed Processing Technology. DOI: http://dx.doi.org/10.1016/B978-0-12-804727-9.00007-7

One major advantage for the top-spray is the capacity of the bowl which can be up to 1500 kg. The production Wurster coating system, on the other hand, is capable of up to 600 kg, and the production size rotary fluid bed module has capacity of about 250 kg. Similarly, a production top-spray system typically has only one nozzle and one pump as compared to the largest Wurster module, which has seven nozzles and ideally requires seven pumps (one for each column), and the rotary granulator/coater which has three or more nozzles and pumps. Thus, fewer variables need to be considered in top-spray coating as opposed to the other methods. Additionally, top-spray units require less clean-up and downtime between batches [1,2].

A major consideration is the product. Nearly all film formers enter a tacky phase before drying which is a leading cause of agglomeration. The separation of small particles by air suspension allows coating with little or no agglomeration, but the finer the substrate, the more difficult discrete coating becomes. The tackiness of some film formers can be reduced by additives such as talc suspended in the coating solution. In general, film formers sprayed from a solution are the most prone to agglomerate particles, whereas latexes (or pseudo-latexes) and materials sprayed from hot melts are likely to be less troublesome. The differences in film quality produced by the three fluid bed coating processes are more pronounced when an organic solvent system is used (as opposed to an aqueous dispersion latex system). The latent heat of vaporization of organic solvents generally is lower than that of water. Thus, spray drying is more problematic for the top-spray process using an organic solvent system rather than an aqueous system.

7.2.1 Hot Melt Coating in Fluid Bed
Top-spray coating is most suitable for hot melt coating because of its ability to operate with the product temperature closest to the congealing temperature of the melt. The selection criteria for coating materials include its substrate release properties in a desired environment, its melting point, melting range and viscosity. Typically, coatings should have a melting point of less than 85°C. The liquid is maintained at a constant temperature during application, which is typically 40−60°C above its melting point. This means that the liquid temperature may be as high as 140−150°C which requires special processing challenges such as storage, delivery and modification of the equipment. A triaxial nozzle setup with a hot atomization air and insulated wand will assure that the molten fluid will be delivered on the fluidized substrate particles [3].

The coating material, melted by heating and sprayed onto the particles, is directly solidified by cold air rather than drying. This confers on hot melt several important production advantages: short processing time, no particle shrinkage, no drying step required, low energy consumption, no solvent used, i.e., low cost, flexible and consistent.

A series of well designed experiments was implemented in a pilot scale unit with 20 kg product capacity to investigate the effects of the process variables on the efficiency of the hot melt coating of Cefuroxime Axetil with stearic acid. Results showed that higher yields with lower operating costs were achieved when the fluidization airflow rate was adjusted by considering the changes in the amount of materials present in the unit as well as the changes in the terminal velocities of particles during the process [4].

7.3 WURSTER COATING

In the Wurster process, the product flow through the apparatus and in and out through the coating zone, is well controlled. The fluidizing gas is fed to the conical bottom of the bed and, under the right conditions, forms a spout or central core in which solids are stripped from the dense annular bed and dragged into the upward moving central gas core where the liquid spray impinges on the solid bed material as it moves through the spray zone. Upon contact with the surface of the particles, the droplets of liquid spread over the surface of the particles and partially coat the solid surface. The repeated motion of the particles through the spray zone allows a continuous coat of material to buildup, resulting in a smooth and, hopefully, uniform coat. The Wurster-based coating process is a complex process with many interrelated concurrent processes. Because of this complex process, much effort should be directed toward the optimization of the process parameters at an early development stage.

Four different regions within the equipment can be identified: the up-bed region, the expansion chamber, the down-bed region, and the horizontal transport region. The size of these regions is determined by the dimensions of the coating apparatus. See Fig. 7.1.

During the coating phase, several processes take place concurrently, i.e., atomization of the film solution/suspension, transport of the film droplets to the substrate, adhesion of the droplets to the substrate, film formation, coating of the substrate, and the drying of the film. In

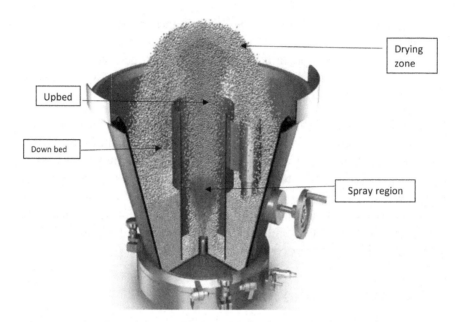

Figure 7.1 Typical Wurster module with different regions that particles go through. Courtesy: IMA, Italy.

coating particles, coalescence must be avoided, and to avoid particle coalescence where colliding particles rebound, the kinetic energy upon particle collision must exceed the viscous dissipation in the liquid and elastic losses in the solid phase [5].

Airflow is critical to the Wurster process. Depending on the particle size, shape, density, and surface area, the airflow must be sufficient to maintain a pressure drop which will move the particles through the coating zone; it may be increased as the coated particles become heavier. The up-bed region is the most difficult to control, because the coating occurs in this region. The product flow in the up-bed region is a dilute vertical pneumatic conveying. The pneumatic conveying is controlled by the up-bed fluidization air rate. Slugging is a frequent problem with the flow in this region for dense and large particles. The particle terminal velocity is dependent on the height of the expansion chamber.

Prewarming the system without the product with approximately 60°C temperature inlet air usually provides a good environment for coating. The exhaust air temperature should be maintained at approximately 40°C, depending on the product and the coating solvent used. The process may be controlled by inlet air temperature or exhaust air temperature.

Wurster process is an excellent process used for solution/suspension layering of drug substance followed by polymer coating to produce modified release pellets.

7.3.1 Operating a Wurster Coater

1. When running a Wurster-based fluid bed coating process, several balances should be kept under control. These are:
   - The product circulation through the apparatus must be adequate, but not too fast. This implies: Up-bed air velocity should be above minimum slugging velocity but not so high as to cause attrition for the product in question. Controlling the fluidization depends on the quantity of product in the coater, inlet airflow, and the column gap. By controlling these three parameters, a proper circulation of the product is assured. The control can be achieved by measuring the pressure drop inside the unit.
   - Expansion chamber and down-bed air velocities should be below minimum fluidization velocity.
2. The spray rate of the coating solution must be adjusted to:
   - The drying capacity in the down-bed region.
   - The product movement in the up-bed region.
   - The lower explosion limit in the up-bed and expansion chamber regions when using solvents.
   - The spray rate is critical. An established equilibrium must be maintained between the application of the coating material and the rate of evaporation or drying. The initiation of spraying should be slow and should be increased to the proper equilibrium after a film has been formed on the particles. A spray rate that is too slow may result in spray drying, nozzle clogging, and lengthening of the process time.
3. The droplet size in the spray must be within limits that secure:
   - Adequate spreading of the coat over the surface of the product being coated.
   - Sufficient slow drying of coating solution before impacting the product to secure proper film formation and avoid spray drying.
4. Equipment set up for Wurster coating is very critical. Assuring following will help avoid processing problems:
   - Check process filter bags (either single or dual shaking) for any tare around the base of the socks or on individual socks.
   - Make sure the filter frame assembly when raised, does not leave any gap and seal is evenly inflated.

- Perform nozzle spray test at the same atomization pressure as required in the process to assure smooth uninterrupted flow.
- Make sure the nozzle(s) are centered in the column so spray does not spray on the walls of the column.
- Wurster insert with multiple nozzles, each nozzle should at the same height inside the column.

As the product is scaled up, the air distributor under the column and down-bed region needs to be selected to obtain optimum movement of particles through the spray zone; one such selection criteria is presented in Table 7.1. Distributor plates are labeled with letters of the alphabet indicating the open areas of the air distributors.

Table 7.1 Guidelines for Air Distributor Selection of Glatt Wurster Coater [6]		
Column	Pellet Sizes (μ)	Distributor Combination
Glatt 6″ Wurster column	<500 Micron	A
	250 << 1200 Micron	B
	600 << 1800 Micron	C
	>1200 Micron	D
Glatt commercial models	<300 Micron	A-I
	150 << 800	B-I
	500 << 1200	B-H
	700 << 1400	C-H
	800 << 1800	C-G
	>1500	D-G

The Wurster technique offers excellent heat and mass transfer within the product bed and can produce uniform coats [7]. However, its efficiency for coating powders has been limited due to the propensity of the fine particles to agglomerate during the coating process [8]. The substrates circulate through various regions of the coater in a fountain-like manner. Each region could contribute to agglomeration if process conditions were not optimal. Since the invention of the Wurster coater, several modifications have been attempted to improve the coating process in bottom-spray air suspension coaters.

7.4 OTHER COATING MODULES

7.4.1 Precision Coater

One of the modifications of the conventional Wurster coater is the Precision coater [9], which uses a modified mode of air distribution to

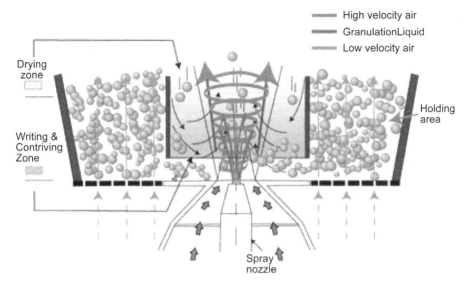

Figure 7.2 Schematic of a precision coater with a swirl accelerator and a flush nozzle to the air distributor. Courtesy: GEA Pharma Systems.

improve the Wurster coating process (Fig. 7.2). It has a swirl accelera-tor under the air distribution plate which swirls and accelerates the processing air to impart spin and high velocity to the substrates as they transit through the partition column. Based on a similar concept employed by Dannelly and Leonard [10] a bicylindrical insert, with a narrower opening at the top, is placed in the central part of the air dis-tribution plate to accelerate the inlet air. This insert is referred to as the swirl accelerator insert.

7.4.2 FlexStream System for Coating

A recent offering by GEA is a system called FlexStream, which can be used for granulating and coating using the same container (Fig. 7.3).

The FlexStream processor is a modified tangential-spray fluidized bed processor with the capability of linear scale-up. In this processor, swirling fluidization airflow is generated by the nonsifting gill plate. The swirling airflow enables a rotational motion about the central axis and it has been shown in various studies to improve heat transfer and increase drying efficiency during the fluidized bed coating pro-cess when compared with nonswirling airflow. Another important feature of the processor is the low-pressure side-spray enveloping air-flow, which is extracted from the inlet plenum and conditioned the same way as the fluidization airflow. This airflow could extend the

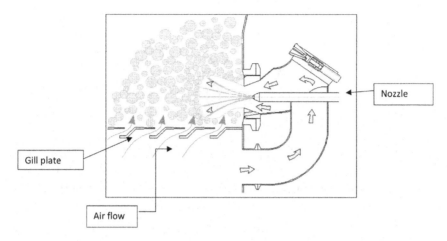

Figure 7.3 FlexStream system with a side entry nozzle. Courtesy: GEA Pharma Systems.

reach of the liquid spray to deeper regions of the particle bed and consequently create a particle-free zone around the spray nozzle, and hence prevent local over-wetting of the core particles by the liquid spray. Xu et al. [11] explored the feasibility of coating irregular-shaped drug particles with wide size distribution using the FlexStream processor module and evaluated the coated particles for their coat uniformity and taste masking efficiency.

7.4.3 Rotary or Tangential Fluid Bed Coating

Producing denser granules was the initial application of tangential or rotary fluid bed processing module. However, now this technique can be used for producing high potency pellets by layering drug onto inert particles by solution or suspension deposition. Subsequently a controlled release coating can be applied using the same equipment setup. This processing technique with its physical principles is quite unlike bottom-spray coating, with the production motion provided by a motor driven rotor disk. Uniform statistical residence time is warranted by defined rotor revolution speed. The coating material is sprayed concurrently inside the rotating product. Different manufacturers have introduced these rotary modules with different designs, e.g., unit with inside wall *(as in Rotoprocessor offered by GEA)* or without inside wall *(as offered by Glatt and other manufacturers) or with a conical rotor module* as offered by Freund-Vector) (Fig. 7.4).

These modules utilize a variable speed rotating disk at the base of the bed. Fluidizing air flows between the edge of the disk and the bed wall

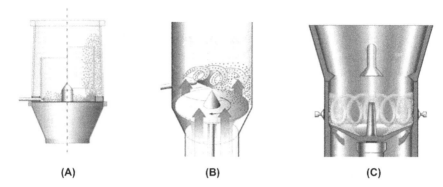

Figure 7.4 Rotary fluid bed modules from GEA (A), Glatt (B), and Vector-Freund (C). Courtesy: GEA Pharma Systems, Glatt Group, and Freund-Vector.

upwards through the bed of solids. The combined action of the fluidizing gas and rotating disk causes the bed solids to circulate. One drawback of tangential-spray coating is the potential for strong mechanical forces during the process. Such forces are beneficial during granulation because they provide good mixing, but they are not desirable during coating because they can cause substrates to break. However, this high kinetic energy makes it somewhat difficult to coat very small particles and is generally destructive for larger and nonspherical products. The film quality achieved via tangential-spraying has been shown to be comparable to that obtained via the bottom-spray process.

A significant advantage of tangential-spraying over top-spray or bottom-spray processes is the option of connecting a powder feeder to minimize exposure of compounds to water or solvent [12]. This technique permits the production of pellets with high-dose loading of actives in a relatively short time. For making pellets using these rotor modules, dry powder can be fed simultaneously with a liquid binder solution. With the correct location of feeds, the substrate is first wetted by the liquid and then covered with a layer of powder. Successive cycles of wetting and powder layering allow the rapid buildup of powder on the substrate solids. This technique is particularly useful when the solute (coating powder) cannot be dissolved or suspended in a liquid or when large amounts of active ingredients must be applied to a substrate. During the drug layering a polymer solution is used as a binder while applying the powdered drug on the substrate particles. High coating quality is shown to be achieved by performing the process close to saturation and spraying droplets small enough to obtain high spraying rate, but not too small to cause incomplete coverage of the core particles [13].

7.5 DRY POWDER COATING

Dry powder coating is an innovative method for the coating of solid dosage forms without the use of organic solvents, and with limited amount of water, if needed, with a very short processing time.

Dry powder coating processes consist of the same sequence of steps that is employed with conventional solvent-based coatings. An excellent review by Sauer et al. [14] highlights dry powder with thermal adhesion coating in fluid bed along with other techniques. During the dry powder coating process, the substrates are often heated above the glass transition temperature of the layering materials so that the coating materials soften and adhere to the substrate. This is followed by the application of coating material to the substrate, relying on the adhesive nature of the formulation to maintain uniformity of coating during the film formation process. Film formation occurs by a process of evaporation, coalescence and sintering which are influenced by process and formulation considerations. The polymer is applied as a micronized powder directly onto the dosage form with or without the simultaneous spraying of a liquid plasticizer. Dry powder coating formulations are engineered to provide the necessary thermal characteristics. Compositionally, this is achieved through the incorporation of polymer, plasticizer, opacifier, colorant and anti-sticking agent. Unlike traditional formulations, higher levels of plasticizers are required to ensure adhesion and film formation. The coating of pellets with powdered polymer offers several advantages, e.g., significant time savings can be accomplished by applying the polymer as a dry powder instead of dissolving it in a liquid to spray it on, thus eliminating organic solvent, producing few agglomerates, higher processing yield, and reduced attrition. Terebesi et al. [15] coated drug loaded pellets with different solubilities. Ethylcellulose (7 and 10 cp viscosity grades) and Eudragit RS were used for dry powder coating of pellets in a fluidized bed coater. Preplasticized ethylcellulose powder was used. Film formation and extended drug release was achieved with ethylcellulose, a polymer with a high glass transition temperature (T_g).

Some of the common problems encountered during fluid bed coating due to changes in formulation and process parameters are listed in Table 7.2.

7.6 PROBLEMS IN COATING IN FLUID BED

Table 7.2 Coating Problems due to Changes in Formulation and Process Parameters

	Parameters	Possible Effect Caused by Change in Parameters
Formulation parameters	Coating liquid viscosity	Increasing the viscosity increases the chance of agglomeration. In addition, increase in viscosity can lead to morphological irregularities. Low viscosity coating solution provides smaller droplet size but if the droplet sizes are too small, spray drying might take place prior to droplet impact on the substrate surface. Too high or too low viscosity may cause poor coating quality.
	Substrate surface prior to coating	For a uniform coating the substrate surface ideally should be smooth and with few pores. Adhesion of the coating liquid is however improved by substrate roughness.
	Coating additives	Additions of surfactants, emulsifiers, anti-adherent agents etc. may reduce the chance of agglomeration but also reduce the mechanical and leaching properties of the final coated product.
	Difference between substrate particle size and droplet size	The chance of agglomeration is reduced as the substrate diameter is increased and the droplet size decreased.
Process parameters	Bed temperature and drying time	If the bed temperature is too high, porous coating layer might result. Chance of agglomeration is minimized with high temperature, unless Methacrylic polymers are used for coating. Amorphous coatings with lower porosities and improved mechanical properties can be achieved by combination of high bed temperature, short drying time and the addition of surfactants to the coating solution.
	Fluidization velocity	Fluidization velocity should be significantly higher than the minimum fluidization velocity (Umf). This will ensure that the average particle circulation time is low enough to minimize agglomeration while obtaining uniform coating.
	Coating flow rate and nozzle pressure	In a typical binary nozzle, pressure through the nozzle governs the flow rate and the droplet size of the coating solution. Typical atomization pressure should be between 0.5 and 3.5 bar. Proper droplet size and proper impact velocity can be achieved by controlling the nozzle atomization pressure. The low droplet velocity may result in uneven coating. The volume of atomization air needs to be sufficient to be able to atomize liquid feed rate.
	Particle circulation time	Particle circulation time should not exceed a few seconds to keep the moisture contents low and reduce the chance of agglomeration. If particle circulation time is too small, shrinkage and cracking in the coating may happen due to inhomogeneous drying of the coated layer.
	Size of spray zone	As the size of the spray zone increases, the variance of the coating distribution decreases.
	Nozzle position	Wurster fluid bed insert with a bottom position of spray nozzle produces more uniform and homogeneous coating layer than the top-spray, especially if the particle sizes are smaller than 125 microns. Also, taller units with larger expansion chamber facilitate proper coating.

7.7 CASE STUDIES
7.7.1 Case Study 1 [16]
Propranolol HCl—loaded pellets (10% w/w drug loading) were prepared by layering a drug-binder solution (21.7% w/w propranolol HCl, 1.0% w/w hydroxypropyl methylcellulose, 0.1% w/w polyethylene glycol, 40.8% w/w ethanol, 36.4% w/w water) onto nonpareil beads in a ball coater (Kugelcoater UNILAB-05, Huttlin, Steinen, Germany). Drug loaded pellets were coated with ethanolic solutions of ethylcellulose, Eudragit L100-55 and blends thereof, or with the respective aqueous polymer dispersions. Drug release from the pellets as well as the mechanical properties, water uptake, and dry weight loss behavior of thin polymeric films were determined in 0.1 M HCl and phosphate buffer, pH 7.4. The researchers concluded that drug release strongly depended on the type of coating technique. Interestingly, not only the slope, but also the shape of the release curves were affected, indicating changes in the underlying drug release mechanisms. The type of coating technique strongly affects the film microstructure and thus, the release mechanism and rate from pellets coated with polymer blends.

7.7.2 Case Study 2 [17]
Phynylpropolamine (PPA) powdered drug, blended with 0.5% w/w of fumed silica was layered onto nonpareil beads using Rotoprocessor [model MP1-Aeromatic (now GEA)] using a powder feeder. Aqueous solution of 4% w/w Hydroxypropylcellulose (HPC) was used a binder after seal coating with 10% HPMC solution. The dried pellets were coated with Surelease solution. The processing conditions for powder layering are listed in Table 7.3 *and film coating conditions are listed in* Table 7.4.

These coated pellets were analyzed for dissolution, porosity, moisture content, particle sizes, morphology, and uniformity of coating using scanning electronic microscopy and powder layering efficiency. The results were satisfactory.

7.8 CHARACTERIZATION OF THE COATING

The relationship between film thickness and size distribution, or any other pellet attribute, is extremely important. The release rate of a drug from the pellets is controlled by the thickness of the film. The coating process in the layering regime leads to finding the optimum process conditions, which lead to the best coating quality. Coating quality can be described in terms of the uniformity of the coating thickness and the

Table 7.3 Processing Conditions for Powder Layering in MP1 [17]

Process Parameters During Powder Layering	Value
Batch size (16/20 sieve cut), g	1000
Target quantity for drug layer, g	500
Rotor speed, rpm	200
Binder spray rate, g/min	4.8 ± 0.2
Powder addition rate, g/min	15 ± 1
Amount of binder sprayed, g	168 ± 8
Total time for layering, min	37 ± 2
Inlet air temperature, °C	25
Outlet air temperature, °C	20−22
Bed temperature, °C	23−14
Gap air pressure, bar	2−3
Atomization air pressure, bar	1
% Yield	90−95

Table 7.4 Process Parameters for Film Coating of Drug-Loaded Pellets in MP1 [17]

Process Parameters During Polymer Coating	Value
Batch size (16/20 sieve cut), g	1000
Target quantity for drug layer, g	500
Rotor speed, rpm	200
Binder spray rate, g/min	10−16
Inlet air temperature, °C	60
Bed temperature, °C	45
Gap air pressure, bar	3
Atomization air pressure, bar	1.5
Fluidizing air volume, cfm	140

porosity of the coating. Using a small droplet size leads to a high spraying rate that is required for obtaining high coating quality. Nevertheless, the droplet size should not be too low, since that may cause incomplete coverage of core particles [13]. The modification of processing parameters such as atomizing air pressure and fluidizing air volume, while slightly reducing the extent of the variation of film thickness in certain cases, can also impact on the quality film formed with respect to surface morphology (monitored by SEM [scanning electron microscopy]), extent of solubilization of core components (monitored by EDX (Energy Dispersive X-Ray Spectroscopy)), and performance (monitored by dissolution). The uniformity in coating thickness can be quantified by the minimum coating thickness and the span of the coating thickness distribution. A wide span of the coating thickness distribution will lead to a big variation in the coating transport properties, which are particularly important for coating applied for controlled/extended release purposes. The presence of film thickness variations and its effect on the dissolution characteristics of the pellets coated in the bottom-spray mode, combined with the potential for segregation, during encapsulation, of the various size particles of a free-flowing pellets system, point to the need to limit the extent of particle size variation of the starting product through optimization of the bead-making process [18–20]. Free-flowing pellets of different sizes often segregate during handling. If the various size pellets were to segregate before or during encapsulation, the potential would exist for the product to vary greatly with respect to release rate, or dissolution characteristics. Therefore, it is important to understand the impact of size distribution, surface area and mass (density) on the film thickness of various size pellets dispersed in a single bed.

7.9 SELECTION OF THE PROCESS FOR COATING

Table 7.5 provides an overview of the three main fluid bed coating processes based on several properties. For both process and economic considerations, the top-spray process is the most desirable. However,

Table 7.5 Comparison of the Three Fluid Bed Coating Techniques [21]					
No	Parameters	Top-Spray	Bottom-Spray	Rotary or Tangential-Spray
	Process considerations			
	Simplicity	3	2	1
	Nozzle access	3	1	2
	Scale-up issues	3	2	1
	Mechanical stress	3	2	1
	Product considerations			
	Surface morphology	1	3	3
	Coating uniformity	2	3	3
	Layering efficiency	1	3	3
	Product coating capacity	2	3	3
	Economic considerations			
	Space	2	1	3
	Production capacity	3	2	1
	Cost	3	2	1
where 1 = easy, 2 = difficult, and 3 = more difficult				

for product applications in which film quality, coating uniformity, and coaling efficiency are important, the bottom-spray and tangential-spray processes are more desirable than the top-spray process [21]. In certain applications, one technique is favored over the other two methods. For example, the top-spray process generally is a good choice for hot melt coating, and bottom-spraying generally is the method of choice for controlled release coating. Tangential-spraying should be considered for producing pellets of high-dose actives.

7.10 SUMMARY

Fluid bed coating offers the possibility to alter and to improve various characteristics of core particles such as the surface properties in a single unit operation. Thre are various ways particles can be coated in fluid bed. The top-spray process is simple and does not require any additional module to coat particles but it has limitations in how small a particle can be coated, and how uniform coating can be applied. Hence, for a controlled release application, top-spray may not be a suitable option. The hot melt coating can be applied using the top-spray, but considerable modification of equipment and solution delivery system is required. For layering solution or suspension and subsequently applying fiunctional coat on the substrate, the Wurster technique has proven to be the choice over the last 30-plus years and the technique is well established in the industry. The rotary processors are used mainly for making pellets. Powder layering is an excellent technique for which the rotary fluid bed module offers a fast way of producing pellets. It is essential, that the coating quality be uniform, which requires that the substrate particles be as uniform in particle size, porosity and particle density as possible. The selection of any of these techniques is dependent upon quality attributes required for the final dosage forms.

REFERENCES

[1] Schaefer T, Worts O. Control of fluidized bed granulation-V. Arch Pharm Chemi, Sci Ed 1978;6:78−81.

[2] Schaefer T, Worts O. Control of fluidized bed granulation-III. Arch Pharm Chemi, Sci Ed 1978;6:12.

[3] Jones DM, Percel PJ. Coating of multiparticulates using molten materials. In: Ghebber-Sallassie I, editor. Multiparticulate oral drug delivery. Marcel Dekker publ; 1994.

[4] Kulah G, Kaya O. Investigation and scale-up of hot-melt coating of pharmaceuticals in fluidized beds. Powder Technol 2011;208:175−84.

[5] Iveson SM, Litster JD, Hapgood K, Ennis BJ. Nucleation, growth and breakage phenomena in agitated wet granulation processes: a review. Powder Technol 2001;117:3—39.

[6] Sonar GS, Rawat SS. Wurster technology: Process variables involved and Scale up science. Innovations Pharm Pharm Technol 2015;1(1):100—9.

[7] Porter SC, Bruno CH. Coating of pharmaceutical solid-dosage forms. In: Lieberman HA, Lachman L, Schwartz JB, editors. Pharmaceutical dosage forms, tablets. New York: Marcel Dekker; 1990. p. 77—159.

[8] Jono K, Ichikawa H, Miyamoto M, Fukumori Y. A review of particulate design for pharmaceutical powders and their production by spouted bed coating. Powder Technol 2000;113:269—77.

[9] Walter K. Apparatus for coating solid particles. United States Patent, 5,718,764; 1998.

[10] Dannelly CC, Leonard CR. Apparatus for spray coating discrete particles. United States Patent, 4,117,801; 1978.

[11] Xu M, Heng PWS, Liew CV. Evaluation of coat uniformity and taste-masking efficiency of irregular-shaped drug particles coated in a modified tangential spray fluidized bed processor. Expert Opin Drug Deliv 2015;12(10).

[12] Othomer DF. Fluidization: background, history and future of fluid bed systems. New York: Reinhold Publishing Corporation; 1956. p. 102—15.

[13] Laksmana FL, Hartman Kok PJA, Vromans H, Frijlink HW, Van der Voort Maarschalk K. Development and application of a process window for achieving high-quality coating in a fluidized bed coating process. AAPS Pharm Sci Tech 2009;10(3).

[14] Sauer D, Cerea M, DiNunzio J, McGinity J. Dry powder coating of pharmaceuticals: a review. Int J Pharm 2013;457(2):488—502.

[15] Terebesi I, Bodmeier R. Optimized process and formulation conditions for extended release dry polymer powder-coated pellets. Eur J Pharm Biopharm 2010;75:63—70.

[16] Lecomte F, Siepmann J, Walther M, MacRae RJ, Bodmeier R. Polymer blends used for the coating of multiparticulates: comparison of aqueous and organic coating techniques. Pharm Res 2004;21(5) (© 2004).

[17] Vuppala M, Parikh DM, Bhagat H, et al. Application of powder layering technology and film coating for manufacture of sustained release pellets using rotary fluid bed processor. Drug Dev Ind Pharm 1997;23(7):687—94.

[18] Wesdyk R, Joshi YM, De Vincentis J, Newman AW, Jain NB. Factors affecting differences in film thickness of beads coated in fluidized bed units. Int J Pharm 1993;93:101—9.

[19] Brown DT. Semiquantitative investigation of tablet coats by electron probe microanalysis. Drug Dev Ind Pharm 1986;12(10):1395—418.

[20] Hossain M, Ayres J. Variables that influence coat integrity in a laboratory spray coater. Pharm Technol 1990;10:72—82.

[21] Srivastava S, Mishra G. Fluid bed technology: overview and parameters for process selection. Int J Pharm Sci Drug Res 2010;2(4):236—46.

CHAPTER 8

Pelletization

8.1 INTRODUCTION

Controlled release preparations can be administered orally in single or multiple unit dosage forms. Among multi-component dosage forms, the most commonly used dosage form is the pellet. It is an ideal dosage form, which on oral administration, disperses or disintegrates rapidly in the stomach to release drug particles, granules and spheroids. Pellets are agglomerates of fine powders or granules of bulk drugs and excipients. They consist of small, free-flowing, spherical or semi-spherical solid units, typically from about 0.5 to 1.5 mm, and are intended usually for oral administration. Polymer-coated multi-particulates or Multiple Unit Pellet System (MUPS) have several therapeutic and technological advantages over single-unit dosage forms. Being small (<2 mm), pellets or multi-particulates can distribute evenly in the gastrointestinal tract, resulting in fewer adverse effects. Pellets also reduce the risk of dose dumping compared to single-unit dosage forms, and result in a reproducible bioavailability [1,2]. Pellets can be either filled into capsules or compressed into tablets as a dosage form to be administered. Pellets offer great flexibility in pharmaceutical solid dosage form design and development. They flow freely and pack easily without significant difficulties, resulting in uniform and reproducible fill weight of capsules and tablets [3–8]. Processing conditions play a significant role in the development of good quality pellets, but the physical forces first bind the primary particles together and initiate the pelletization process. Physical forces determine the nature of the elementary growth mechanism that forms the basis of pellet formation in any processing equipment. There are numerous publications on the theory of formation of a pellet [9–13].

Several techniques are available to manufacture pellets that contain uniform drug distribution. The most widely used pelletization unit operation is the extrusion/spheronization. Other techniques include high shear pelletization, melt pelletization, spray congealing, etc. The

How to Optimize Fluid Bed Processing Technology. DOI: http://dx.doi.org/10.1016/B978-0-12-804727-9.00008-9

scope of this chapter will cover pelletization using the fluid bed processor with its different processing modules. Drug layering involves depositing the drug on the surface of a substrate. The layering process comprises the deposition of successive layers of drug entities from solution, suspension, or dry powder on nuclei, which may be crystals or granules of the same material or inert starter seeds. Pellets are produced in the fluid bed from powders using a rotary fluid bed module, and by solution/suspension or powder layering onto inert spheres such as nonpareils or microcrystalline cellulose. The most common way of depositing the drug on the substrate is by using air-suspension coating using the Wurster coating module.

In solution/suspension layering, drug particles are dissolved or suspended in the binding liquid. By controlling the rate of solution/suspension application, the agglomeration of substrate particles can be minimized. In powder drug layering, a binder solution is first sprayed onto inert seeds, followed by the addition of drug powder. The solution layering in the fluid bed processor can be performed either by top-spray, bottom-spray or by using a rotary processor unit in the fluid bed. Pellets are ideal for modified release application due to their spherical shape and a low surface area-to-volume ratio. Coated pellets composed of different drugs can be blended and formulated in a single dosage form. This approach facilitates the delivery of two or more drugs, chemically compatible or incompatible, at the same sites or different sites in the gastrointestinal tract. Even pellets with different release rates of the same drug can be supplied in a single dosage form.

8.2 PELLETIZATION WITH WURSTER COATING MODULE

Wurster coaters are bottom-spray fluid bed coaters and are used for layering solution or suspension of drug, besides applying functional coatings on the pellets. The sections of the Wurster module have three distinct regions. The nozzle area inside the column, where particles are sucked by the air current to the entry, the up-bed region where liquid is applied on the particles called the spray region, and the down-bed where particles exit from the column and dry as they fall back into the down-bed region. The spray nozzle supplies liquid solution as well as an atomization air that breaks the solution into small droplets and forms the spray region in the column. As the pellets travel through this region, they receive a coating solution in the form of droplets, and as they move around in the fluidized bed, the liquid droplets are dried by the hot airflow to form a layer on each substrate particle. As the layered particles exit the column into the expansion area of the processor, the velocity of these particles slows down while undergoing drying. The dried pellets then reenter the down-bed region for additional coating of solution or suspension (see Fig. 8.1). The area of the air distributor plate directly under the column has more perforated area than the periphery region of the air distributor, resulting in a higher central air velocity through the center column per nozzle. The air flow in the down-bed region keeps material in suspended form and draws horizontally into the gap at the base of the column. The height of the column, the gap between the column and the air distributor, and the air velocity through the spray zone control the rate of substrate flow horizontally into the coating zone. During layering in progress, mass increases gradually so the height of the column may be adjusted to maintain the smooth transfer of pellets in the coating zone. Layering or superposition of different layers of droplets around the particle results in a homogeneous reservoir system. After several cycles of wetting-drying, a continuous film is formed, with a controlled thickness and composition depending on the materials used. It is mainly at this stage that the tendency for agglomeration between two or among several particles is high (Fig. 8.2).

Table 8.1 lists some of the variables during solution/suspension layering.

Figure 8.1 Schematic of Wurster and coating container with bottom nozzle. Courtesy: GEA Pharma Systems.

Figure 8.2 Suspension layered pellet showing substrate sugar sphere (smooth) at the center surrounded by drug layer.

Table 8.1 Variables for Wurster Coating/Solution-Suspension Layering	
Variables	
Equipment	• Wurster column diameter and height • Gap between the column and the air distributor • Nozzle (number, air cap settings, port diameter) • Pump type • Distance between pump and nozzle input
Layering solution/suspension	• Solids contents, viscosity • Solution or suspension • Pump • Solution/suspension hold time • Uniformity of solution/suspension throughout the process
Process	• Inlet air temperature and dew point • Airflow • Atomization air pressure • Spray rate • Product quantity • Spray time • Drying time

8.3 PELLETIZATION IN ROTARY FLUID BED

8.3.1 Pelletization by Agglomeration

Alternative techniques for producing pellets are the single-pot methods, where pellets are produced, dried, and coated in the same equipment. They are -step processes that take place in one machine, such as a high-sheer mixer or a rotary fluid bed processor. Using one machine for the whole process ensures batch-to-batch reproducibility and reduction of production time and cost, and enables automation of the process. The rotary fluid bed module for the fluid bed was originally developed to perform granulation processes and was later expanded to perform other unit operations including the manufacturing and coating of MUPS and is now used for pelletization by powder layering and solution/suspension layering as well as polymer coating. An advantage of rotary fluid bed processing to produce granules over conventional top spray granulation technique was reported by Jager and Bauer [14]. In this unit, the conventional air distributor is replaced by the rotating disk. The material to be pelletized is loaded on the rotating disk. The binder solution is added through the atomization nozzle located tangentially to the wall of the bowl. The centrifugal force creates a dense, helical doughnut-shaped pattern. This type of motion is caused by the three directional forces. The vertical movement is caused by the gap or slit air around the rotating disk; the gravitational force folds back the material to the center, and the centrifugal force caused by the rotating disk pushes the material away from the center. The granulation/pellets produced in the rotary fluid bed processor show less porosity compared to the conventionally agglomerated product in the fluid bed processor. The 1972 patent [15] for the rotor technology was awarded for the equipment and coating of the granular material. The subsequent patents [16,17] were awarded for the coating of the spherical granules. Different manufacturers of these units offer different designs of the rotor module but the principle of motion in the units is similar. See Figs. 8.3−8.6.

The double-wall unit (Fig. 8.4) is more or less similar to tangential spray or centrifugal fluid bed granulator in terms of principle and design. It is double-walled and the process is carried out with the inner wall in the open or closed position. The inner wall is closed, so that simultaneous application of liquid and powder could proceed until the MUPS have reached the desired size. The inner wall is then raised and the spheres enter into the drying zone. The pellets are lifted by the

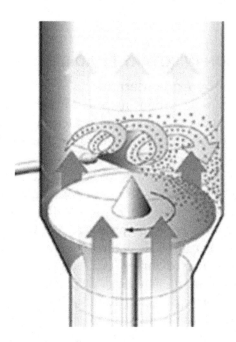

Figure 8.3 Rotor insert. Courtesy: The Galtt Group.

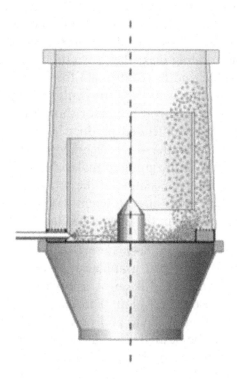

Figure 8.4 Two-walled "rotoprocessor" insert. Courtesy: GEA Pharma Systems.

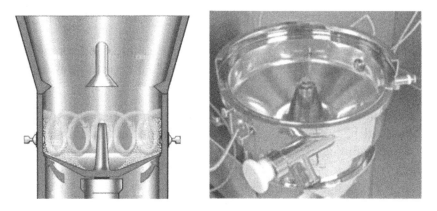

Figure 8.5 Freund-Vector Granurex (GXR) rotoprocessor. Courtesy: Freund-Vector.

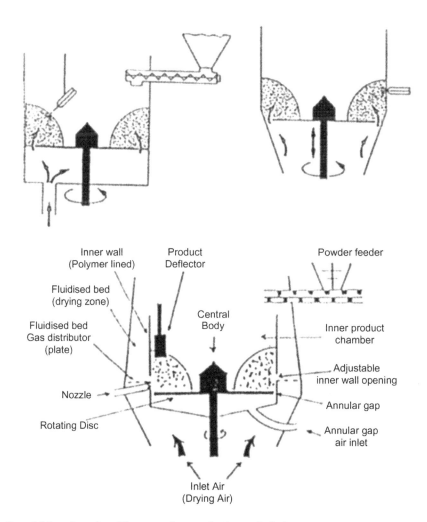

Figure 8.6 Rotor insert from different manufacturers showing powder feeding.

fluidization air up and over the inner wall back into the forming zone. The cycle is repeated until the desired residual moisture level in the pellet is achieved.

Korakianiti et al. [18] studied the preparation of pellets using the rotary fluid bed. Authors concluded that the rotor speed and amount of water significantly affected the geometric mean diameter of the pellets, and they proposed an equation to show that correlation. Pišek et al. [19] studied the influence of rotational speed and surface of rotating disk on pellets produced by using a rotary fluid bed. They used a mixture of pentoxifylline and microcrystalline cellulose to produce pellets using suspension of Eudragit NE 30 D as a binder. The results showed that both surface and rotational speed of the disk have influence on shape, surface and size of the pellets while there was less effect on the density, humidity content and yield. They found the textured surface of the disk-produced pellets with a rougher surface when rotational speed was increased compared to the smooth surface, where increased rotational speed produced more spherical pellets with larger diameters. Kristensen and Hansen [20] compared granulation prepared in the fluid bed with a top-spray and rotary processor and concluded that the rotary processor offers better maneuverability in terms of the obtainable granule size and was less influenced by the flow properties of the starting materials. Iyer et al. [21] conducted experiments where solution containing phenylpropanolamine (PPA) HCl and a binder was sprayed on nonpareils by top-spray and rotary-spray. They concluded that pellets made by the rotary-spray technique were superior to those made by the top-spray technique in terms of surface smoothness, yield, and uniformity of drug content.

Bed moisture content is a critical property. The three main factors that affect the bed moisture content are the air flow, air temperature and binder spray rate. A low bed moisture content leads to coarse granules that are very porous, since they grow by a crushing and layering mechanism. A reduction of the air flow rate or gap air pressure decreases the helicoidal flow pattern in the bed [22]. An increase in air gap pressure can induce a higher moisture loss, which contributes to a reduction of the particle size. Environmental and other processing variables that might have impact on the process should be controlled. That is also applicable for the formulation being processed. By controlling these external variables, residual variations will be lower, resulting in more accurate analysis of variance. Addition of more water or binder solution will produce larger pellets but it may produce pellets with wider particle size distribution (PSD).

8.3.2 Pelletization by Layering Solution or Suspension

Layering is a slow growth mechanism and involves the successive addition of fragments and fines on an already formed nucleus. Layering on cellulose or sugar [nonpareil] spheres is a process that can lead to the uniform distribution of a drug on a carrier excipient. The fluid layering technique can be useful for active agents or intermediates dispersed in liquids. In the layering step, the number of particles remains the same, but the total mass in the system increases due to the increasing particle size as a function of time. The rotary fluid bed is used for producing a pellet by layering the active drug suspension or solution onto nonpareil cores and subsequently coating them with polymers to impart modified release properties [23]. Baki et al. [24] reported the importance of surface free energy to understand the layering process. The spreading coefficient plays an important role in adhesion of the layered drug solution on the substrate. The authors prepared the carrier pellets from mixtures of a hydrophilic (microcrystalline cellulose) and a hydrophobic (magnesium stearate) component in different ratios. These cores were coated in a fluid bed apparatus with an aqueous solution of an active agent, with or without the addition of hydroxypropyl methylcellulose (HPMC) as an adhesive component. The connection between the core and the layer and its mechanical properties was evaluated extensively, varying the composition of the core. The authors noted that the effectiveness of the layering of less hydrophobic core can be increased through the application of a binder. The peeling of the layers containing HPMC was observed for hydrophobic cores, but this phenomenon was not detected for the hydrophilic ones. Hileman et al. [25] reported manufacture of immediate spheres of a poorly water-soluble drug in a rotary fluid bed by layering the active drug suspension onto nonpareil cores. These immediate release spheres were then over-coated with an ethylcellulose/HPMC hydroalcoholic solution in the same unit eliminating the need for additional process and handling steps. Mise et al. [26] investigated the effect of particle size and the wettability of API on drug layered pellets for three different APIs with high drug load (>97) formulation. Acetaminophen (APAP); Ibuprofen (IBU); and Ethenzamide (ETZ) were used as model drugs based on their differences in wettability and PSD. Pellets with mean particle sizes of 100−200 μm and a narrow PSD could be prepared regardless of the drug used. IBU and ETZ granules showed a higher sphericity than APAP granules, while APAP and ETZ granules exhibited higher granule strength than IBU. The relationship between drug and granule properties suggested that the wettability and the PSD of the drugs were critical parameters affecting sphericity and granule strength, respectively. Furthermore, the dissolution profiles of granules prepared with poorly water-soluble drugs (IBU and ETZ) showed a rapid release (80% release in 20 min) because of the improved wettability with granulation.

8.3.3 Pelletization by Powder Layering

Various equipment manufacturers have promoted powder layering on the pellets in a rotary fluid bed. See Fig. 8.6. In 1992, Jones et al. received a patent for such a process [27]. The process claims to have advantages of layering a drug substance with relatively a small amount of liquid, thus making this layering process more efficient. Binders most commonly used in drug layering are gelatin, povidone, carboxymethyl cellulose and hydroxypropyl methylcellulose. A typical process involves a rotary fluid bed with tangential nozzle(s) placement. Nonpareils or MCC spheres are loaded inside the rotor insert. Powdered API is fed into the machine via a precision powder feeder. The binder solution binds active powder to the outside of the nonpareils, and the powder begins to coat the substrate. The size of the starting MCC or nonpareils will depend on the amount of drug that will be layered on the pellets. Higher drug loads will require smaller starting substrate particles, but may also pose problems of agglomeration instead of layering. Hence the spray rate, slit air and inlet air temperature are the major factors that will have to be monitored. The most critical factors are the binder spray rate, powder feed rate and the inlet air flow. A powdered drug that does not get an opportunity to stick on the wet substrate will get carried away in the exhaust air, affecting the product yield and the potency. The powdered API particle size and flow properties will have to be appropriate for powder layering. Normally a micronized API is recommended and an augur feeder to meter in the API is required. For best results powder particle size should be <30 microns.

- Several advantages have been mentioned in the literature of powder layering:
  - Significant time saving over solution/suspension coating is accomplished by adding API in powder form rather than by dissolving or suspending into a liquid and spraying on the core pellets.
  - High drug load is achievable compared to solution/suspension layering.
  - Ability to create a multi-layer bead with multiple actives in the same bead by sequentially loading different drugs with a proper barrier between them.
  - Reduces or eliminates the use of organic solvents.
  - Use of inert cores as a starting substrate or crystals of the drug can be layered on.
  - Sphericity is obtained at the end of the process on the pellets due to centrifugal forces in the rotor.
- Some of the disadvantages mentioned in the literature are:
  - The drug loading is limited by the size of the inert core substrate.
  - Binder spray rate and powder feed rate require close monitoring throughout the process.
  - If the API has very poor flow property, additional excipient may have to be added to enhance the flow, thus reducing the API drug loading.
  - Yield losses are in the 10%−20% ranges due to loss of fines in the air stream.
  - Delivery of the powder needs to be as close to the wet pellets as possible to reduce loss of powder.

8.4 CASE STUDIES

8.4.1 Preparation of Immediate Release Pellets With Centrifugal Pelletization [28]

Equipment used: Aeromatic (now GEA Pharma Systems) Rotoprocessor for MP-1

A mixture of Acetaminophen and microcrystalline cellulose was pelletized using **Polyvinylpyrrolidone** (PVP) solution and compared the one with gelatin solution as a binder. Authors performed 2^3 factorial experiment with one replicated center point for each of the two binders PVP and Gelatin. Table 8.2 summarizes 10 experiments each for PVP and Gelatin.

Table 8.2 Case Study Formulation for Immediate Release Pellets With Centrifugal Pelletization in Rotary Fluid Bed

	PVP Binder Solution	Gelatin Binder Solution
Batch size, kg	0.526	0.526
Acetaminophen, %	47.5	47.5
Avicel, %	47.5	47.5
PVP, %	5	0
Gelatin, %	0	5
Water, %	15–30.8	20–30.9
Rotor speed, rpm	798–1204	799–1207
Spray rate, g/min	32.6–68.2	33.4–71.6
Forming time, min	4.5–10	5.08–10.5
Drying time, min	20–33.8	14.8–38.0
Binder solution temperature, °C	22–36	45–60
Room temperature, °C	19.5–22	20–22
Relative humidity, %	70–80	70–78

After characterizing the pellets and comparing them to the ones produced from the extrusion/spheronization process, authors concluded that the rotary processor can produce immediate release pellets of acetaminophen of similar characteristics. The addition of a binder solution had an influence on dependent variables such as usable yield and pellet crushing strength. Pellet crushing strength was improved by using a higher spray rate. Higher rotor speed also increased crushing strength and yield; with lower rotor speed usable yield increased. Determining the end point of pellet growth is critical in this process.

8.4.2 Solution Layering in Rotary Fluid Bed [23]
Aeromatic Rotoprocessor was used and the processing conditions are listed in Table 8.3.

8.4.3 Powder Layering Powder Layering Case Study [3]
Experiments were performed with 1000 g of nonpareil as substrate and Phenylpropanolamine (PPA) as a model drug for powder layering in an Aeromatic (now GEA Pharma Systems) MP1 Rotoprocessor. The PPA was blended with 0.5% w/w fumed silica in a plastic bag and transferred into the powder feeder. A 4% w/w aqueous solution of Hydroxypropyl cellulose was used as a binder and sprayed with a 1 mm port schlick nozzle. The processing parameters are listed in Table 8.4.

Table 8.3 Process Parameters for Solution Layering in Rotary Fluid Bed

Quantity of sugar spheres 250–300 μ, g	1000
Water-soluble API 64% with PVP binder and water, g	1036
Nozzle tip diameter, mm	1
Number of nozzles	1
Atomization air, bar	1.0
Solution spray rate (slow/fast), g/min	20–60
Rotary disk setting, rpm	500–510
Inlet air temperature °C	80
Exhaust air temperature °C	40–45
Air volume, cfm	180–200
Total process time, min	30

Table 8.4 Process Parameters for Powder Layering in Rotary Fluid Bed

Processing Parameters	Value
Batch size (nonpareils), g	1000
Target quantity of drug layer, g	500
Rotor speed, rpm	200
Binder spray rate, g/min	4.8 ± 0.2
Powder addition rate, g/min	15 ± 1
Amount of binder sprayed, g	168 ± 8
Total time for spraying, min	37 ± 2
Inlet air temperature, °C	25
Outlet air temperature, °C	20–22
Bed temperature, °C	23–14 (*dropped temperature as powder layer progressed*)
Gap air pressure, bar	2–3
Atomization air pressure, bar	1
Yield (%) calculated after processing	90–95

The pellets were seal coated in the same unit with HPMC to a 1% weight gain with a liquid addition rate adjusted to maintain a bed temperature of 45°C. These seal coated pellets were film coated in the same unit with Surelease (diluted to 15%w/w). Pellets were evaluated by Scanning Electron Microscopy, dissolution studies, and porosity measurements. Authors concluded that high loading drug pellets can be prepared by powder layering and subsequently coated in the same equipment.

8.5 CHARACTERIZATION OF PELLETS

Regardless of which manufacturing process is used, pellets have to meet the following requirements:

- The pellets should be evaluated after drying to the specific moisture content.
- They should be near spherical and have a smooth surface, both considered optimum characteristics for subsequent film coating.
The pellets are mainly coated for esthetic, taste masking, stability, enteric-release or controlled release purposes. The coating thickness of the pellets must be uniform in order to achieve any of these end product performances. For uniform coating thickness, the formulation, equipment and process variables are usually selected based on the reproducibility of the size distribution, surface area, shape, surface roughness, density and friability, including the reproducibility of morphologic properties of the pellets.
- The particle size range should be as narrow as possible. The optimum size of pellets for pharmaceutical use is considered to be between 600 and 1000 μm. Uniform size distribution and friability are two of the most physical attributes pellets must have. Normally all pellets are coated with a polymer. For film coating, a narrow size distribution of pellets is a prerequisite in addition to spherical shape and smooth surface. The size distribution affects both the performance of the coating and the release rate of the drug. Pellets should be robust, low porosity, nonfriable with higher density and thus capable of handling during coating, sieving, and encapsulation step without loss of drug from the surface.
- Shape of the pellets can be evaluated with three shape factors as reported by Vertommen [29] as circularity, roundness, and elongation. The purpose of these parameters is to emphasize the aspect ratio of pellets, which should be 1 for a perfectly round pellet.
- The pellets should contain as much as possible of the active ingredient to keep the size of the final dosage form within reasonable limits.
- There are numerous publications detailing methods to characterize the pellets [30–34].

8.6 SUMMARY

Pellets are multi-particulate dosage forms which can be manufactured by using different fluid bed modules. The pellets can be prepared using a mixture of API and the excipients with an addition of binder solution in a rotary fluid bed processing module manufactured by different equipment suppliers. The wet granulation process employed is similar in most of these processing modules from different suppliers. The advantage of a rotary processor is that it offers one module for making, drying and coating pellets as well as layering solution/suspension or layering powder of API. The Wurster module is used for layering of solution/suspension of API and/or coating pellets, and it is the most widely used technique in the industry for commercial production. The understanding of formulation and process variables is essential for a successful outcome. The formed pellets by any of these techniques should be of narrow particle size, robust and of the size that can be further used in encapsulation or compression unit operations.

REFERENCES

[1] Bodmeier R. Tableting of coated pellets. Eur J Pharm Biopharm 1997;43:1−8.

[2] Celik M. Compaction of multiparticulate oral dosage forms. New Jersey: Marcel Dekker, inc.; 1994.

[3] Vuppala MK, Parikh DM, Bhagat HR. Application of powder-layering technology and film coating for manufacture of sustained-release pellets using a rotary fluid bed processor. Drug Dev Ind Pharm 1997;23(7):687−94.

[4] Conine JW, Hadley HR. Small solid pharmaceutical spheres. Drug Cosmet Ind 1970; 90:38−41.

[5] Lyne CW, Johnston HG. The selection of pelletizers. Powder Technol 1981;29:211−16.

[6] Ghebre-Sellassie I, Gordon R, Fawzi MB, Nesbitt RU. Evaluation of a high-speed pelletization process and equipment. Drug Dev Ind Pharm 1985;11:1523−41.

[7] Reynolds AD. A new technique for the production of spherical particles. Manuf Chem 1970;June:39−43.

[8] Niskanen M. Powder layering and coating in a centrifugal granulator: Effect of binder-solution concentration, powder particle size and coating amount on pellets properties. Doctor's Thesis. University of Helsinki; 1992.

[9] Juslin L. Measurement of droplet size distribution and spray angle of pneumatic nozzle, and granule growth kinetics and properties of lactose, glucose and mannitol granules made in a fluidized bed granulator. Doctor's thesis. University of Helsinki; 1997.

[10] Ghebre-Sellassie I. Pellets: a general overview. In: Ghebre-Sellassie I, editor. Pharmaceutical pelletization technology, vol. 37. New York: Marcel Dekker, Inc.; 1989. p. 1−13.

[11] Goodhart. Centrifugal granulators. In: Ghebre-Sellassie I, editor. Pharmaceutical pelletization technology, vol. 37. New York: Marcel Dekker, Inc.; 1989. p. 101−20.

[12] Sastry KVS, Fuerstenau DW. Mechanism of agglomerates growth in green pelletization. Powder Technol 1973;7:97−105.

[13] Sastry KVS, Fuerstenau DW. Kinetic and process analysis of the agglomeration of particulate materials by green pelletization. In: Saatry KVS, editor. Agglomeration 77. AIME; 1973. p. 381−402.

[14] Jäger KF, Bauer KH. Effect of material motion on agglomeration in the rotary fluidized-bed granulator. Drugs Made Ger 1982;25:61–5.

[15] US Patent 3,671,296 (June 20, 1977).

[16] US Patent 4,034,126 (July 5,1977).

[17] US Patent 4,542,043 (September 17, 1985).

[18] Korakianti ES, Rekkas DM, Dallas PP, Choulis NH. Optimization of the pelletization process in a fluid bed rotor granulator using experimental design. AAPS Pharm Sci Tech 2000;1(4) article 35.

[19] Pišek R, Planinšek O, Tuš M, Srčič S. Influence of rotational speed and surface of rotating disc on pellets produced by direct rotor pelletization. Pharm Ind 2000;62:312–19.

[20] Kristensen J, Hansen VW. Wet granulation in rotary processor and fluid bed: comaprison of granule and tablet properties. AAPS PharmSciTech 2006;7 article 22 PPE1-E10.

[21] Iyer RM, Augusburger LL, Parikh DM. Evaluation of drug layering and coating: effect of process mode and binder level. Drug Dev Ind Pharm 1993;19:981–9980.

[22] Ghebre-Sellasie I, Knoch A. 2nd ed. Encyclopedia of pharmaceutical technology, vol. 3. New York: Marcel Dekker Inc.; 2002.

[23] Parikh D.M. "Layering in rotary fluid bed a unique process for the production of spherical pellets for controlled release" Presented at Interphex-USA, New York, NY. May 18, 1991.

[24] Baki G, Bajdik J, Djuric D, Knop K, Kleinebudde P, Pintye-Hodi K. Role of surface free energy and spreading coefficient in the formulation of active agent-layered pellets. Eur J Pharm Biopharm 2010;74:324–31.

[25] Hileman G.A., Sarabia R.E. "Manufacture of immediate and controlled release spheres in a single unit using fluid bed rotor insert" Presented at the Annual Meeting of the American Association of Pharmaceutical Scientists (AAPS), Poster PT 6167, 1992.

[26] Mise R, Iwao Y, Kimura S, Osugi Y, Noguchi S, Itai S. Investigation of physicochemical drug properties to prepare fine globular granules composed of only drug substance in fluidized bed rotor granulation. Chem Pharm Bull 2015;63:1070–5.

[27] US Patent 5, 132, and 142 (1992).

[28] Robinson, Hollenbeck. Manufacture of spherical acetaminophen pellets: comparison with rotary processor with multiple steps extrusion spheronization. Pharm Technol 1991;48–58.

[29] Vertommen J. "Pelletization in a rotary processor using the wet granulation technique" Thesis, Acta Biomedica Lovaniensia, Leuven University Press; 1998.

[30] Wan LSC, Lai WF. Factors affecting drug release from drug-coated granules prepared by fluidized bed coating. Int J Pharm 1991;72:163–74.

[31] Sarisuta N, Sirithunyalug J. Release rate of indomethacin from coated granules. Drug Dev Ind Pharm 1988;14(5):683–7.

[32] Cahyadi C, Sheng Koh JJ, Loh ZH, Chan LW, Sia Heng PW. A feasibility study on pellet coating using a high-speed quasi-continuous coater. AAPS PharmSciTech 2012;13(4) (# 2012)

[33] Rieck C, Bück A, Tsotsas E. Stochastic modelling of particle coating in fluidized beds The 7th World Congress on Particle Technology (WCPT7) Procedia Eng 2015;102:996–1005.

[34] Pawar AS, Bageshwar DV, Khanvilkar VV, Kadam VJ. Advances in pharmaceutical coatings. Int J Chem Tech Res 2010;2(1):733–7.

CHAPTER *9*

Other Fluid Bed Processes and Applications

9.1 INTRODUCTION

The fluid bed process, over the last 50 years, has transformed from a simple drying unit operation to agglomeration, coating and various applied technologies. Notable among them are those using low melt waxes and polymers to granulate in a fluid bed, or spraying melted polymer to spray on the powders through a heat traced hot air atomization setup, using steam to affect agglomeration, pressure swing granulation (PSG), instantizing of a product for making fast-dissolving granules, taste masking of bitter drugs, etc. The fluid bed process offers advantages in making the desired product by manipulation of formulation or process parameters to produce modified release forms, improve the solubility, modify the density, and mask the bitter taste for oral solid dosage products among other applications.

9.2 FLUIDIZED HOT MELT GRANULATION (FHMG)

Melt granulation belongs to the group of hot-melt technologies which represents an alternative to the classic solvent-mediated technological processes of agglomeration. Melt granulation is an emerging technique based on the use of binders that have relatively low melting points (between 50 and 80°C) and act as a molten binding liquid, or combining low melting binders with powders and using the fluid bed hot air to melt the binders that effectively act as liquid binders to form granules. Application of this approach can also produce solid dispersion of poorly soluble drugs to increase the solubility and bioavailability. The main advantage of hot-melt processes is the absence of solvents, which can be efficiently utilized in enhancing chemical stability of moisture-sensitive drugs and also improving their physical properties. Moreover, the drying phase is eliminated, which results in a more economical and environmentally friendly process. There are also some limitations in using melt granulation processes. Melt granulation, or thermoplastic granulation, is based on agglomeration carried out by means of a

How to Optimize Fluid Bed Processing Technology. DOI: http://dx.doi.org/10.1016/B978-0-12-804727-9.00009-0

binder material, which is solid at room temperature, but softens and melts at higher temperature (i.e., 50–90°C). When melted, the action of the binder liquid is similar to that in a wet-granulation process. The binder is added either in a powder form to the starting material at ambient temperature, followed by heating the binder above its melting point (in situ granulation), or in a molten form, sprayed on the heated materials in the fluid bed (*spray-on* granulation).

During in *situ granulation,* the melting material and other powders are added to the fluidized bed and the inlet temperature is set to an appropriate value to start granulation by melting the binder. The system is cooled to a solid state with the molten excipient acting as a binder. For *spray-on* granulation, the meltable binder is heated, and the liquid binder is maintained above the melting point of the binder until it is sprayed by a two-fluid nozzle. The tubing carrying the molten binder needs to be heat traced (heated) as well, as atomization air is at a high temperature to avoid cooling the binder as it exits the nozzle. When the granulation end point is reached, the material in a fluidized bed can be rapidly cooled via the fluidizing air. In either case, the process temperature should be adjusted to avoid binder solidification prior to particle/particle collision. The probability of particle/binder collision can be increased by adjusting the binder addition rate and the powder flux through the spray zone. In addition to the binder amount, agglomerate growth during fluidized hot melt granulation (FHMG) is also determined by the viscosity of the molten binders [1–3]. Overall the most important critical variables determining the particle size and granule quality during FHMG are concentration, viscosity, spray rate and droplet/particle size of the binder, primary particle size, bed temperature, atomization pressure, air velocity, and atomization pressure. However, fluid bed melt granulation is less sensitive to the air volume used to fluidize the powder bed compared to wet granulation. During wet granulation the rate of liquid evaporation is determined by the airflow rate, and it is essential that the spray rate is sufficient to exceed the drying rate induced by the airflow to ensure agglomeration and consolidation of the particles during processing. Since no evaporation of the liquid phase occurs during melt processing, the effect of this variable is eliminated. The major drawback is the required high temperature during the process, which can cause degradation and/or oxidative instability of the ingredients, especially of thermolabile drugs. Binder viscosity was shown to significantly influence granule growth; it may be concluded that low viscosity, highly mobile molten polymers will produce a system that grows via coalescence and layering, whereas higher viscosity binders will result in a system that fails to grow after powder layering has occurred because of the inability of the binder to migrate to the outer surface of the colliding granules [4]. Kukeca et al. [5] showed that melt granulation using hydrophilic binders is an effective method to improve the dissolution rate of a poorly water-soluble drug. The binder addition procedure was found to influence the dissolution profile obtained from granules produced in FHMG. The spray-on procedure resulted in a higher dissolution rate of carvedilol from the granules. FHMG has been proposed as an approach to taste masking bitter drugs [6]. Waxy binders have been used in the preparation of conventional and sustained release tablets [7], and more recently in the preparation of fast-release tablets [8]. Yanze et al. [9] reported preparation of effervescent granules using polyethylene glycol (PEG) 6000 as a melt binder using the fluidized bed melt granulation. The melt solidification technique for the preparation of sustained release ibuprofen beads with cetyl alcohol has been studied in the laboratory [10]. Similarly, other waxy carriers like beeswax, carnauba wax, microcrystalline wax, Precirol ATO5, Gelucire 64/02 have been studied [11,12] in the preparation of microspheres. Sustained release ibuprofen mini-tablets have been prepared by the melt extrusion technique using microcrystalline wax and starch derivatives [13]. Tablets with a shorter disintegration time than 10 min were obtained with 2.0% crospovidone added as a disintegrant. In comparison to tablets prepared from the wet-granulated mass, employment of the FHMG method resulted in tablets with faster dissolution of carbamazepine (more than 80% of the drug released within 15 min). This was achieved with mannitol or lactose/microcrystalline cellulose (MCC), as fillers [14]. Zai et al. [15] prepared gastroretentive extended-release floating granules using the FHMG process. The floating granules exhibited sustained release exceeding 10 h. The drug release and floating properties can be controlled by modification of the ratio or physical characteristics of the excipients used in the formulation. The stability of enalapril maleate was improved by FHMG [16]. The granules obtained showed adequate flowability and a fast dissolution rate of enalapril maleate with almost 100% of the drug released in 10 min.

Table 9.1 shows the common meltable binders used for melt granulation.

Table 9.1 Common Meltable Binders for Fluidized Hot-Melt Granulation

Common Meltable Binders Used in Melt Granulation

Hydrophilic Meltable Binder	Melting Point °C	*Hydrophobic* Meltable Binder	Melting Point °C
Gelucire 50/13	35–44	Beeswax	56–60
Poloxamer 188	~50.9	Carnauba wax	75–83
Polyethylene glycol 2000	42–53	Cetyl palmitate	47–50
Polyethylene glycol 3000	48–63	Glyceryl behenate	67–75
Polyethylene glycol 6000	49–63	Glyceryl monostearate	47–63
Polyethylene glycol 8000	54–63	Glyceryl palmitostearate	48–57
Polyethylene glycol 20,000	53–66	Glyceryl stearate	54–63
Stearate 6000 WL1644	46–58	Hydrogenated castor oil	62–86
		Microcrystalline wax	58–72
		Paraffin wax	47–65
		Stearic acid	46–69
		Stearic alcohol	56–60

9.2.1 Case Study

Ingredients: Ibuprofen [2-(4-isobutylphenyl)-propionic acid] (Ibu) and ketoprofen [(3-benzoylphenyl)-propionic acid] as model drug-micronized

Excipient-Lactose 200 mesh and PEG 6000 as a meltable binder

Batch size 200 g (Table 9.2).

Table 9.2 Case Study

Fluidized Hot Melt Granulation Parameters

Process	Inlet Airflow Rate (m³/h)	Inlet Air Temperature (°C)	Product Temperature	Time(min.)
Mixing	16.6	25	25.0–30.0	5
Heating	23.9	80	25.0	3
Cooling time	239	25	58.0–40.0	3

Technological Properties of Granules Prepared From FHMG

API	Yield (%)	Bulk Density (g/cm³)	Tap Density (g/cm³)	Carr's Index (%)	Friability (%)
Ibuprofen	86	0.464	0.497	6.7	0.73 ± 0.16
Ketoprofen	96	0.555	0.558	5.6	0.89 ± 0.09

Source: *Modified From Passerini N, Calogerà G, Albertini B, Rodriguez L. Melt granulation of pharmaceutical powders: a comparison of high-shear mixer and fluidised bed processes. Int J Pharm 2010;391: 177–186, [17].*

9.3 SPRAY CONGEALING

This process, also known as spray chilling or spray cooling, is a melt granulation technique whereby solid particles are dissolved or dispersed in a molten carrier (having a melting point above room temperature). The resulting liquid is atomized in a chamber at a temperature below the melting point of the carrier (i.e., often at ambient temperature) and solid spherical microparticles are rapidly formed when the molten carrier in the droplets congeals upon contact with the cooler airflow in the spray congealing apparatus. The spray congealing process involves spraying a hot melt of wax, fatty acid, or glyceride into an air chamber below the melting point of the meltable materials or at cryogenic temperature. Upon cooling, particles of $10-3000\ \mu m$ in diameter are obtained. As there is an absence of solvent evaporation, the congealed particles are strong and nonporous. Based on its operating principle, spray congealing relates to spray drying as conventional melt granulation relates to wet granulation. Depending on the carrier material this technique can prepare sustained release particles for oral drug delivery [18], enhance the dissolution rate of poorly soluble drugs (even without the formation of solid dispersions [19−20]) and mask the taste of a drug [21].

The major limitation of spray congealing, however, is the difficulty in achieving drug loads above 25%, as the formulation must be pumped towards the nozzle and atomized in the cooling chamber. At higher drug loads the molten formulation is often too viscous to obtain nonaggregated spherical particles having a narrow particle size distribution after atomization through a pneumatic two-fluid, rotary (or centrifugal) or ultrasound-assisted nozzle. A pressure nozzle is less often used for spray congealing, as the high viscosity of the formulation requires too high a pressure.

From a processing point of view the dimensions of the cooling chamber must be sufficient to ensure a high process yield, hence a large spray congealing tower may be required when a high production capacity or large particles are required. In addition, the nozzle type used for atomization also determines the dimensions of the cooling chamber depending on the trajectory of the particles; a rotary nozzle requires a wider chamber, whereas a two-fluid nozzle can be combined with a higher but narrower chamber to optimize process yield. Hence spray congealing is more economical to produce by using a spray dryer.

9.4 PRESSURE SWING GRANULATION (PSG)

PSG uses pressure to agglomerate powders without any binders in a rotating fluidized bed. In PSG, the powder bed is fluidized in one cycle and compacted in the other cycle. This facilitates the formation of spherical granules without any binders. Dry coating, which attaches tiny submicron-sized particles (guests) onto relatively larger micron-sized particles (hosts) without using any solvent, binders or even water, is a promising alternative approach.

The PSG technique utilizes the spontaneous agglomeration nature of cohesive fine powders or Geldart's group C powders. In the PSG fluidized bed, fine powders are granulated by cyclic fluidization and compaction, conducted by alternating upward and downward gas flow. During the compaction period, a sharp spike-like pressure reversal created by opening a valve of the upper pressure chamber returns the fine particles collected by the bag-filter to the bed, and at the same time destructs channels in the bed. The compaction process and the granulation-fluidization process are repeated by turns. In the compaction process, the gas flows from the upper side to the lower side, and the powder bed is wholly compacted. In the granulation-fluidization process, air bubbles rise through the powder bed and the powder particles are forced close together and compacted in the particle-thick domain formed under the air bubbles. As these processes are repeated, the granules are formed mainly by the van der Waals force of the particles. The PSG method enables the manufacture of granules that consist of pure drug particles. Fitrah et al. [22] used PSG to produce ibuprofen and lactose granules in the ratio of 3:7 utilizing gas at 70°C, eliminating the sticking tendency during tableting. Masayuki Watanabe et al. [23] prepared agglomerates of milled sodium salicylate powders (sodium salicylate and calcium gluconate), granulated by the PSG method, for producing granules for an inhalation application.

9.5 WURSTER MODULE FOR GRANULATION

Top-spray fluidized bed granulation is the current gold standard in the industry. The bottom-spray fluidization technique, alternatively termed as Wurster processing, has been used primarily for pellet coating and less so for granulation. In the Wurster process, the granulating liquid is atomized and sprayed directly onto particles supported and suspended by an upwardly moving air stream. Based on the Wurster process, Ichikawa and Fukumori [24] proposed the concept of micro-agglomeration, a technique where pulverized powders are converted into agglomerates $20-50 \, \mu m$ in size for subsequent microencapsulation by film coating. Rajniak et al. [25] granulated Mannitol, A-tab and Avicel with 15 % binder solution to produce granules with more uniform drug distribution and tighter particle size distribution. The Wurster process is also used to produce pellets by layering solution or suspension of the active pharmaceutical ingredient (API) on an inert core (microcrystalline cellulose or sugar spheres, etc.) or crystal of the API, which then can be coated with polymer for modified release application.

One of the modifications of the Wurster process is called a precision granulator, which uses a modified mode of air distribution to improve the fluid dynamics of the system [26]. A high velocity air stream, which is also rotating, is established in the central tube. Particles are picked up at the base of the tube and accelerated by the air stream. While in the stream, the particles are contacted with liquid droplets produced by the spray nozzle at the base of the tube. Within the central tube the relative velocities of air, liquid droplets and particles are high, so wetting is efficient and drying begins almost immediately. The agglomerates are dry by the time they leave the top of the tube. The materials are not fluidized; they are pneumatically conveyed by the air stream. The air velocity is therefore not as critical as in a fluid bed device. A highly ordered particle circulation pattern and unique fluid dynamics in precision granulation were advantageous for the wet granulation of moisture-sensitive and low-dose drugs [27,28]. Liew and co-workers [29] demonstrated the suitability of precision granulation for materials that are soluble, sticky or hygroscopic in nature. They compared the properties of granules produced from precision granulation, top-spray granulation and high-shear granulation on an industrial scale and found that the porosity, strength and density of granules produced from precision granulation were intermediate between those produced from top-spray and high-shear granulation processes. At equivalent tablet weight and hardness, tablets produced from precision granulation exhibited shorter disintegration times.

9.6 FOAM GRANULATION

Foam granulation technology was developed by the Dow Chemical Company. This technology enables faster granulation, which involves a simpler wet granulation processing of materials and employs a high-shear mixer, or fluid bed processor. Foam technology involves creating the foam of the binder such as Hypromellose and adding that to the powders to granulate. Foamed binders have a high spread-to-soak ratio. Particles are coated rather than soaked, making binder distribution more consistent and placing the binder where it is needed: on the surface. Since there is no nozzle involved, it eliminates all the variables associated with nozzles (nozzle configuration, distance from moving powder bed, nozzle clogging, spray patterns, droplet size and distribution, etc.). Less water is required to agglomerate and evaporate and there is a shorter production time [30–35]. Research and application of a foam binder in high-shear granulation have been increasing in the industry, but application in fluid bed granulation is not as widespread yet.

9.7 STEAM GRANULATION

This technology is a simple modification of the conventional wet granulation method in which steam is used as binder instead of water, and involves the injection of a jet of steam into the bed of fluidized particles to be granulated. Steam at its pure form is transparent gas, and provides a higher diffusion rate into the powder and a more favorable thermal balance during the drying step. After condensation of the steam, water forms a hot thin film on the powder particles, requiring only a small amount of extra energy for its elimination, and evaporates more easily. The advantages of this process include the higher ability of the steam to distribute uniformly and diffuse into the powder particles, production of spherical granules with larger surface area resulting in an increase in dissolution rate of the drug from granules [36–43], and favorable thermal balance resulting in rapid drying and shorter processing time. However, the process has some disadvantages as well; key among them is that it is not suitable for thermolibe APIs, and not suitable for all binders, requires special equipment for steam generation and its transportation, and since the steam temperature is about 150°C, local overheating and excessive wetting of the particles in the vicinity of the nozzle could create lumps. Sotome et al. [44] used a steam/water two-step technique to granulate food powder in a fluid bed. Superheated steam and water (127°C, 138 kPaG) were sprayed at 18.8 g/min and 0~40 g/min respectively through a single-fluid nozzle to the mixed powder of corn starch (800 g) and dextrin (200 g). The amount of water sprayed to the powder decreased to 40%~84% to produce the granules having equivalent size with that obtained by the conventional fluidized bed granulation using the liquid binder.

9.8 ROTARY FLUID BED AGGLOMERATION

Rotor technology is a "single-pot" method of pellet and denser granule production, based on fluid bed technology. There are different designs produced by different equipment manufacturers but the basic concept is the same. The perforated air distributor of the conventional fluidized bed is replaced by a rotating solid metal disk. By raising and lowering the disk it is possible to alter the space between the disk's edge and the wall of the container, and the airflow can be adjusted. The materials to be processed in a rotor module are subjected to a combined force of centrifugation and fluidization. Rotor technology has been used to produce denser granules. The use of a rotor processor in the production of spheroids had been reported by Robinson and Hollenbeck [45]. In a single-step procedure, the whole operation of spheroid formation, drying and coating may be confined to a single piece of equipment. Spheroids are formed directly from materials in the powder form by an agglomeration-spheronization process. The powder handling capacity requires a modification of the spheronizer chamber. A positive pressure must be maintained between the annular gap to prevent powder slippage between the frictional plate and the rotor housing. Among the advantages of rotary fluid bed are better mixing and coating of particles, and the potential for obtaining spherical beads and granules from powders. During spraying of the binder solution, the flow is tangential and co-current, leading to a more homogeneous material distribution.

The surface and rotational speed of the disk have influence on shape, surface and size of pellets while there is less effect on true density, humidity content and yield of the experiment. Keeping rotational speed of the smooth disk constant during agglomeration of powder particles and increasing rotational speed during spheronization of agglomerates result in more spherical pellets with larger diameters and smoother surfaces.

Agglomeration in a rotary processor using solutions of Polyethyleneglycol (PEG) as the primary binder liquid was found to be a robust process [46]. The process can be applied for the preparation of agglomerates with a high drug content and physical properties suitable for further processing, such as coating, capsule filling, or compression. The process allowed for the incorporation of up to 42.5% wt/wt PEG and can therefore be used as an alternative to melt agglomeration with hydrophilic meltable binders.

Another big advantage of the rotary fluid bed is the capability of powder layering on the substrate particles. The rotor will need some modification to feed the powder in a consistent manner to the rotating particles or pellets, and the binder solution is sprayed tangentially. As the pellets become wet, the powder sticks to the pellets, and because of the fluid bed, they dry rapidly. Caroline Désirée Kablitz et al. [47] developed a process in a rotary fluid bed with a three-way nozzle and a gravimetrical powder feeder achieving optimum coating material application, to obtain an enteric film avoiding completely the use of organic solvents and water. Using hydroxypropyl methylcellulose acetate succinate, an enteric coat was obtained without adding talc as an anti-tacking agent. The coating process was a highly efficient process with short processing time of 23 min coating phase and 45 min curing phase causing low energy consumption.

9.9 SPRAY-DRYING GRANULATION

A spray dryer process involves atomizing a pumpable liquid, a solution, a suspension or slurry. A typical spray dryer consists of a chamber, atomizing wheel or spray nozzle, pump, and source of heated gas (to dry the millions of droplets formed), a dry powder collection container, a cyclone to separate the air and fines escaping the unit, and a secondary filter to capture fines or in some cases a scrubber in case the organic solvent is used as a carrier liquid. The spray dryer is configured as either a closed or open cycle unit depending on whether the solution/suspension contains organic solvent or water. It is a continuous process where a dry product is obtained by feeding a solution or a suspension of active agent with or without excipients to the drying system, where the feed is atomized and dried with a heated gas stream followed by subsequent separation of powder/granular product from the gas stream.

Spray drying is a unique process compared to other granulation methods in several ways. The solution/suspension is a homogeneous liquid, which results in uniform distribution of all components. Many granulation methods employ mechanical energy to produce granules while in spray drying; the product is subjected to the shear forces unlike in the high-shear granulation process. When nozzles or centrifugal atomizers are used, the energy imparted from these does not adversely affect the formed particles. The active drug is never in contact with any moving parts within the spray dryer; hence cleaning issues are minimized [48].

Simplicity of the spray-drying process and its ability to manipulate the particle size of the resultant powders have opened up the possibility of using spray-drying technology to produce directly compressible "drum-to-hopper" granulation. Ibuprofen and acetaminophen granulations were manufactured using a spray dryer with formulations containing drug, microcrystalline cellulose and other excipients in aqueous suspension; the resultant granulation showed desirable flow and compression properties when compressed in tablets [49]. By employing a spray dryer, a number of unit operations to produce granulation for tablets or capsules were eliminated. Modification of crystal properties of acetazolamide to improve compression behavior has been reported [50].

9.10 CONTINUOUS GRANULATION

Continuous processing has long received support from the FDA (US Food and Drug Administration), which has been interacting with the European Medicines Agency, and aligns well with the FDA's process analytical technology (PAT) and process validation guidance. The term "continuous" is applied to all production or manufacturing processes that run with a continuous flow. With that definition, continuous processing of solid dosage products in the pharmaceutical industry means starting the process from the synthesis of API to the final packaging of tablets or capsules 24/7 all year round. In the 1980s, Koblitz and Ehrhardt [51] reported on continuous wet granulation and drying. The article focused on continuous variable frequency fluid bed drying. Berkovitch in a Manufacturing Chemist article [52] quoted some researchers presenting these concepts in a symposium. Continuous processing of pharmaceuticals, including a process for solid oral dosage form manufacturing, was also discussed by Kawamura [53].

Drivers for continuous manufacturing in the pharmaceutical industry include the fact that New Chemical Entities are getting more potent/toxic. Small, dedicated suites are suitable to serve as "containments" for highly potent drugs. Investment costs for multi-product facilities for highly potent oral solid dosage products are exaggerated due to segregation measures. FDA initiatives and the recent availability of compact integrated systems from equipment and software suppliers continue to drive the process. Perceived operational and labor savings for high volume products can be a significant driver.

Continuous granulation in a fluid bed is more common where large quantities of product are required. The current interest in continuous granulation focuses on a combination of high shear/twin-screw extruders coupled with a continuous dryer. Following are some of the systems offered by the equipment suppliers.

GEA Pharma Systems offers the "Consigma" integrated granulation/drying and tableting system. Based on the twin-screw wet granulation and the fluid bed drying, this integrated system is capable of producing between 0.5 and 200 kg of product depending on the size of the system. In twin-screw continuous wet granulation, powdered solids and binder pass through a twin-screw extruder, usually with corotating screws. The resulting granulated product is dried continuously, milled, blended and compressed.

Many unit operations are intrinsically continuous and are well understood. For all remaining unit operations, equipment is available. Experiences with continuous wet granulation have been positive. Opportunities to adopt a continuous process exist and should be built on the quality by design approach, which will require more advanced control systems with simple and more complex PATs. It must be realized that not all products or processes will be manufactured with continuous granulation approach. Every API will have to be evaluated for its capability to be a candidate for continuous granulation and a pertinent process will have to be developed for it.

Bohle offers the BCG System with screw configuration of the twin screw. Granulation uses a twin- screw extruder, and drying is performed using infrared and vacuum. The throughput of 8–30 kg/h is controlled by a dosing unit. The residence time varies according to the throughput required and product attributes desired.

Glatt Pharma Systems offers a continuous granulation system in which the product is granulated in a continuous fluid bed granulator/dryer. The composition to be granulated is fed at one end of the unit, and the binder liquid is added to the product as it travels through the fluid bed unit, continuous granulating and drying until dry product is discharged.

Loedige offers a continuous granulation system in which a premixed mixture of composition is fed in the system, and by a very high rotation speed the product is moved through the horizontal drum as ring layer. Liquid addition is done by injectors or through the hollow shaft. The granulated product is then fed into a continuous fluid bed dryer with a varying fluidization velocity as the product is transferred through the dryer and further processed as required.

As of this writing the US FDA has approved two products utilizing continuous manufacturing. However, having the process understanding and robust risk management program with continuous manufacturing will be the key to their successful implementation. The continuous manufacturing approach must be seen as the platform for process development that is enhanced by continuous manufacturing. Continuous manufacturing provides renewed attention to PAT, process control and on-line instrumentation. Implementation of continuous processing will be based on how manufacturers will overcome the challenges of compliance, process design, process control and ultimately quality.

REFERENCES

[1] Abberger T. Influence of binder properties, method of addition, powder type and operating conditions on fluid-bed melt granulation and resulting tablet properties. Pharmazie 2001;56:949–52.

[2] Seo A, Holm P, Kristensen HG, Schaefer T. The preparation of agglomerates containing solid dispersions of diazepam by melt agglomeration in a high shear mixer. Int J Pharm 2003;259:161–71.

[3] Wong TW, Cheong WS, Heng WS. Melt granulation and pelletization. In: Parikh DM, editor. Handbook of pharmaceutical granulation technology. New York: Informa Health; 2005. p. 385–406.

[4] G.P. Andrews, D.S. Jones, G.M. Walker, S. Bell, M.A. Vann, Pharmaceutical Technology Europe, 01 June 2007.

[5] Kukeca S, et al. Characterization of agglomerated carvedilol by hot-melt processes in a fluid bed and high shear granulator. Int J Pharm 2012;430:74–85.

[6] Kidokoro M, Haramiishi Y, Sagasaki S, Yamamoto Y. Application of fluidized hot-melt granulation (FHMG) for the preparation of granules for tableting. Drug Dev Ind Pharm 2002;28:67–76.

[7]   Jones D, Percel P. Coating of multiparticulates using molten materials: formulation and process consideration. In: Ghebre-Sellassie I, editor. Multiparticulate oral drug delivery. 1st ed. New York: Marcel Dekker; 1994. p. 113–42.

[8]   Perissutti B, Rubessa F, Moneghini M, Voinovich D. Formulation design of carbamazepine fast-release tablets prepared by melt granulation technique. Int J Pharm 2003;256:53–63.

[9]   Yanze F, Duru C, Jacob M. A process to produce effervescent tablets: fluidized bed dryer melt granulation. Drug Dev Ind Pharm 2000;26:1167–76.

[10] Maheshwari M, Ketkar A, Chauhan B, Patil V, Paradkar A. Preparation and characterization of ibuprofen–cetyl alcohol beads by melt solidification technique: effect of variables. Int J Pharm 2003;261:57–67.

[11] Adeyeye C, Price J. Development and evaluation of sustained release ibuprofen-wax microspheres. I. Effect of formulation variables on physical characteristics. Pharm Res 1991;8:1377–83.

[12] Bodmeier R, Wang J, Bhagwatwar H. Process and formulation variables in the preparation of wax microparticles by a melt dispersion technique for water-insoluble drugs. J Microencapsul 1992;9:89–98.

[13] De Brabander C, Vervaet C, Görtz J, Remon J, Berlo J. Bioavailability of ibuprofen from matrix mini-tablets based on a mixture of starch and microcrystalline wax. Int J Pharm 2000;208:81–6.

[14] Kraciuk R, Sznitowska M. Effect of different excipients on the physical characteristics of granules and tablets with carbamazepine prepared with polyethylene glycol 6000 by fluidized hot-melt granulation (FHMG). AAPS PharmSciTech 2011;12(4) (# 2011). http://dx.doi.org/ 10.1208/s12249-011-9698-7.

[15] Zhai H, Jones DS, McCoy CP, Madi AM, Tian Y, Andrews GP. Gastroretentive extended-release floating granules prepared using a novel fluidized hot melt granulation (FHMG) technique. Mol Pharm 2014;11(10):3471–83. http://dx.doi.org/10.1021/mp500242q Publication Date (Web): August 08, 2014.

[16] Guimarães TF, Comelli ACC, Tacón LA, Cunha TA, Marreto RN, Freitas LAP. Fluidized bed hot melt granulation with hydrophilic materials improves enalapril maleate stability. AAPS PharmSciTech 2016; Aug 3. Epub 2016 Aug 3.

[17] Passerini N, Calogerà G, Albertini B, Rodriguez L. Melt granulation of pharmaceutical powders: a comparison of high-shear mixer and fluidised bed processes. Int J Pharm 2010;391:177–86.

[18] Van Melkebeke B, Vermeulen B, Vervaet C, et al. Melt granulation using a twin-screw extruder: a case study. Int J Pharm 2006;326:89–93.

[19] Li LC, Zhu LH, Song JF, et al. Effect of solid state transition on the physical stability of suspensions containing bupivacaine lipid microparticles. Pharm Dev Technol 2005; 10:309–18.

[20] Maschke A, Becker C, Eyrich D, et al. Development of a spray congealing process for the preparation of insulin-loaded lipid microparticles and characterization thereof. Eur J Pharm Biopharm 2007;65:175–87.

[21] Cavallari C, Luppi B, Di Pietra AM, et al. Enhanced release of indomethacin from PVP/stearic acid microcapsules prepared coupling co-freeze-drying and ultrasound assisted spray-congealing process. Pharm Res 2007;24:521–9.

[22] Bakar NFA, Mujumdar A, Urabe S, Takano K, Nishii K, Horio M. Improvement of sticking tendency of granules during tabletting process by pressure swing granulation. Powder Technol 2007;176:137–47.

[23] Watanabe M, Ozeki T, Shibata T, Murakoshi H, Takashima Y, Yuasa H, et al. Effect of shape of sodium salicylate particles on physical property and in vitro aerosol performance of granules prepared by pressure swing granulation method. AAPS Pharm Sci Tech 2003;4(4) Article 64.

[24] Ichikawa H, Fukumori Y. Microagglomeration of pulverized pharmaceutical powders using the Wurster process I. Preparation of highly drug-incorporated, subsieve-sized core particles for subsequent microencapsulation by film-coating. Int J Pharm 1999; 180(2):195−210.

[25] Rajniak P, Stepanek F, Dhanasekharan K, Fan R, Mancinelli C, Chern RT. A combined experimental and computational study of wet granulation in a Wurster fluid bed granulator. Powder Technol 2009;189:190−201.

[26] Walter KT. Inventor. Precision granulation. US6492024B1; 2002.

[27] Liew CV, Er DZL, Heng PWS. Air-dictated bottom spray process: impact of fluid dynamics on granule growth and morphology. Drug Dev Ind Pharm 2009;35(7):866−76.

[28] Er DZL, Liew CV, Heng PWS. Layered growth with bottom-spray granulation for spray deposition of drug. Int J Pharm 2009;377(1−2):16−24.

[29] Liew C, Walter K, Wigmore A, et al. Precision granulation as an alternative granulation method. Presented at: AAPS poster presentation. Toronto, Ontario, Canada, 2002.

[30] Sheskey P, Keary C, Clark D, Balwinski K. Scale-up trials of foam-granulation technology-high shear. Pharmaceutical Technology Europe 2007.

[31] Sheskey P, Keary C, Inbasekaran P, Deyarmond V, Balwinski K. Foam technology: the development of a novel technique for the delivery of aqueous binder systems in high-shear and fluid-bed wet-granulation applications. Poster presented at the Annual Meeting and Exposition of the American Association of Pharmaceutical Scientists, Salt Lake City, Utah, October 26−30, 2003.

[32] Paul J, Shesky R, Colin K. New foam binder technology from Dow improves granulation process. Pharmaceutical Canada 2006; 19−22.

[33] Thompson MR, Weatherley S, Pukadyil RN, Sheskey PJ. Foam granulation: new developments in pharmaceutical solid oral dosage forms using twin screw extrusion machinery. Drug Dev Ind Pharm 2012;38(7):771−84.

[34] Paul S, Colin K, Dan C, Karen B. Scale-up trials of foam-granulation technology—high shear. Pharm Technol Eur 2007;19(9):37.

[35] Hapgood KP, Litster JD, Biggs SR, Howes T. Drop penetration into porous powder beds. J Colloid Interface Sci 2002;253(2):353−66.

[36] Rodriguez L, Cavallari C, Passerini N, Albertini B, Gonzalez-Rodriguez M, Fini A. Preparation and characterization by morphological analysis of diclofenac/PEG 4000 granules obtained using three different techniques. Int J Pharm 2002;242:285−9.

[37] Cavallari C, Albertini B, Gonzalez-Rodriguez ML, Rodriguez L. Improved dissolution behaviour of steam-granulated piroxicam. Eur J Pharm Biopharm 2002;54:65−73.

[38] Albertini B, Cavallari C, Passerini N, Gonzalez-Rodriguez ML, Rodriguez L. Evaluation of beta-lactose, PVP K12 and PVP K90 as excipients to prepare piroxicam granules using two wet granulation techniques. Eur J Pharm Biopharm 2003;56:479−87.

[39] Vialpando M, Albertini B, Passerini N, Bergers D, Rombaut P, Martens JA, et al. Agglomeration of mesoporous silica by melt and steam granulation Part I: a comparison between disordered and ordered mesoporous silica. J Pharm Sci 2013;102:3966−77. Available from: http://dx.doi.org/10.1002/jps.23700.

[40] Vialpando M, Albertini B, Passerini N, Vander Heyden Y, Rombaut P, Martens JA, et al. Agglomeration of mesoporous silica by melt and steam granulation Part II: screening of steam granulation process variables using a factorial design. J Pharm Sci 2013;102:3978−86. Available from: http://dx.doi.org/10.1002/jps.23699.

[41] Hammer K. Steam granulation apparatus and method. United States Patent 4,489,504, 1984.

[42] Hampel R, Heinrich S, Mörl L, Peglow M. Modeling and experimental analysis of superheated steam granulation. Proceedings of the 5th World Congress on Particle Technology, Orlando/Florida 2006.

[43] Solanki HK, Basuri T, Thakkar JH, Patel CA. Recent advances in granulation technology. Int J Pharm Sci Rev Res 2010;5(3):48−54.

[44] Sotome I, Inoue T, Katagiri T, Takeuchi H, Tsuda M, Okadome H, et al. Reduction of water addition in fluidized bed granulation by steam-water tow-phase binder. JSFI J 2014;15(1):25−35.

[45] Robinson RL, Hollenbeck RG. Pharm Tech 1991;15:48.

[46] Kristensen J. Investigation of a 2-step agglomeration process performed in a rotary processor using polyethylene glycol solutions as the primary binder liquid. AAPS PharmSciTech 2006;7(4) Article 89 (http://www.aapspharmscitech.org).

[47] Kablitz CD, Harder K, Urbanetz NA. Dry coating in a rotary fluid bed. Eur J Pharm Sci 2006;27:212−19.

[48] Loh ZH, Er DZ, Chan LW, Liew CV, Heng PW. Spray granulation for drug formulation. Expert Opin Drug Delivery 2011;8(12):1645−61.

[49] Parikh DM, Erkoboni D. "Spray Drying" presented at Annual AAPS meeting, 2000.

[50] Martino PD, Scoppa M, Joiris E, Palmieri G, Andres C, Pourcelo Y, et al. The spray drying of acetazolamide as method to modify crystal properties and to improve compression behavior. Int J Pharm 2001;213:209−21.

[51] Koblitz T, Ehrhardt L. Continuous variable-frequency fluid bed drying of pharmaceutical granulationspharmaceutical technology, March 1985.

[52] Berkovitch I. From batch to continuous pharmaceutical engineering. Manufacturing Chemist, August 1986, 43−45.

[53] Kawamura K. Continuous processing of pharmaceuticals. In: Swarbrick J, Boyle J, editors. Encyclopedia of pharmaceutical technology. 3rd ed. New York: Marcel Dekker; 1990.

CHAPTER 10

Process Control and PAT

10.1 BACKGROUND

The development of process control in the chemical and pharmaceutical industry started with simple mechanical controllers for temperature and pressure. These were superseded by analog transmitters and controllers—initially with pneumatic, later on with electrical energy supply. Process computers introduced digital control into process plants and were used as open loop systems mainly for recording and process optimization. Hardware of process control systems is based on microprocessor equipped devices; control functions are performed by software.

Optimization of the process starts with successful development of a robust process during the development of product. The Design of Experiments (DoE) does offer learning tools towards that objective. With DoE, critical process variables can be identified. Understanding the responses from experimentation, one can define operating limits of the process as it gets scaled up. Process variables found to be the most significant during the DoE may still apply to the pilot scale and ultimately for the commercial batches. Assessing the results from these experiments requires process analytical tools to learn about the robustness of experiments. Variability in raw materials, equipment, and processing conditions are unavoidable and will cause variability in product quality. A more robust strategy is to develop processes with measurement and control capabilities that compensate for process variability and foster a culture of continuous process improvements.

10.2 PROCESS CONTROL

In production, it is essential to maintain a high level of automation with integrated feedback loops to exercise process controls in real-time. A PID (proportional integral derivative) controller is a commonly used feedback mechanism in industrial control systems. A PID controller

How to Optimize Fluid Bed Processing Technology. DOI: http://dx.doi.org/10.1016/B978-0-12-804727-9.00010-7

attempts to correct the error between a measured and a desired value for a process variable by calculating and then utilizing a corrective adjustment action. In a fluid bed processor, airflow rate and temperature are typical process variables that can be adjusted by a PID controller. If the PID controller parameters are selected improperly, the controlled process input can be unstable. Therefore, the PID loop should be tuned to each application.

According to the 2004 PAT guidance by USFDA [1] the design and optimization of drug formulations and manufacturing processes can include four steps:

1. Identify and measure critical material and process attributes relating to product quality
2. Design a process measurement system to allow real-time or near real-time (e.g., on-, in-, or at-line) monitoring of all critical attributes
3. Design process controls that provide adjustments to ensure control of all critical attributes
4. Develop mathematical relationships between product quality attributes and measurements of critical material and process attributes.

According to ICH Q 9 guidance, quality risk assessment identifies fluid bed granulation as one of the critical unit operations. Particle size distribution, density, flow properties, compressibility and homogeneity are required for characterizing the final granulation. Process measurements can be taken one of three ways: at-line, where the sample is removed and analyzed close to the process stream; on-line, where the sample is diverted from the manufacturing process to an analyzer and possibly returned to the stream; and in-line, an invasive or noninvasive process that analyzes the sample while it is part of the process stream.

The detection of granule growth during fluidized bed processing and the endpoint is traditionally based on the measurements of process parameters. Control of the formulation components and the process is essential to ensure the consistent production of granules with the desired quality characteristics. During the solution spraying period, critical data related to the inlet air humidity (dew point) and temperature, product and outlet air temperature, air flow, binder spray rate, atomizing air pressure, and pressure drops across the bed and filter are monitored. During drying, the inlet air temperature and humidity, product and exhaust air temperature, pressure drops and air flow are continuously monitored. The product and exhaust air temperatures indicate the constant rate and falling rate of drying periods. Process troubleshooting relies on collected data. The type of data and the collection interval vary by vendor and user, but the preference is to collect as much as possible and as frequently as possible. For a fluid bed processor, this would minimally include all temperatures (dew point, inlet, product, and exhaust), process air volume, atomizing air pressure and volume and spray rate. Dependent variables should also be collected— product and outlet filter differential pressure, liquid line pressure. In systems with complex air handling units (AHU), total air volume, preheater temperature, ambient air dew point and dehumidification dew point may be added to the list.

10.3 TEMPERATURE MEASUREMENT

During granulation, the product temperature and exhaust temperatures are monitored. Product temperature will drop as the binder liquid is sprayed, and a steady state is obtained when evaporation and addition of the binder liquid are in perfect balance. An imbalance will be indicated by observing the product or exhaust temperature.

During the drying stage the temperature of the product within the bed or the air leaving the fluidized bed dryer can be measured. The air and product temperatures are cooled as a result of the evaporation of water from the granules. The drying end-point can therefore be estimated by an increase in the air or product temperatures. The moisture content of the granules within the bed is estimated from the difference in temperature between the dry bulb temperature measured within the bed and the wet bulb temperature of the bed when the granules are very wet.

Establishing a drying curve is another method, if the fluid bed processor does not have any other instrumentation for determining the drying end-point. As the drying enters the critical moisture stage, the temperature of the product in the bowl will start increasing. By taking samples from that point every few minutes, and measuring the loss on drying and the corresponding product and exhaust temperature, a correlation can be established and a drying curve can be drawn.

By repeating this procedure, the exhaust temperature range indicating required moisture specification for the dried product is established. This exhaust temperature should be used to determine the end of the drying step.

Another temperature-based method for monitoring drying involves estimating the humidity of the air leaving the dryer by comparing the wet bulb and the dry bulb temperatures. Initially, the humidity of the air leaving the dryer is very high due to evaporation of water from the granules during drying. As the granules dry, the humidity of the outlet air decreases; the end-point of drying occurs when the humidity of the inlet and outlet air approach the same value. Another disadvantage of temperature-based monitoring methods is limited accuracy due to poor fluidization conditions within the bed. This method of detecting the drying end-point via the exhaust air, or product temperature is only repeatable, if the humidity level of the inlet air is controlled.

10.4 NIR

Near-infrared (NIR) spectroscopy has been used to monitor drying by measuring the moisture content of the air moving through the dryer and, more commonly, by measuring the moisture content of the granules within the bed. NIR spectra is an effective alternative to traditional methods, such as thermo-gravimetry and Karl Fischer titration for both water content and water binding determinations. This is because O–H bands of water are very intensive in the NIR region, exhibiting five absorption maxima, the positioning of which depends on the hydrogen bonding intensity. The specific band to be used for water determinations depends on the desired sensitivity and selectivity level. NIR quantification of moisture content is usually an easy task with respect to data processing. Particle size measurements with Near Infrared Spectroscopy (NIRS) in diffuse reflectance mode rely on the particle size dependent scatter effect of powders resulting in nonlinearly sloping baselines [2,3].

NIR can be applied for both quantitative analysis of water and for determining the state of water in solid material. Frake et al. [4] demonstrated the use of NIR for in-line analysis of the moisture content in 0.05–0.07 mm pellets during spray granulation in a fluid bed processor. Fouling of the measurement window by the moist mass challenges the reliable collection of NIR spectra during fluid bed granulation. To minimize the effect of wet mass or dried powder from interfering, the measurement use of an air supply around the sight glass has been used in some applications.

Rantanen et al. [5,6] described a moisture content measurement approach using a rationing of 3−4 selected wavelengths. He and his co-workers reported that the critical part of an in-line process was the sight glass for probe positioning that was continuously blown with heated air. They also reported spectra baselines caused by particle size and refractive properties of the in-line samples. Solvents other than water have also been evaluated for real-time quantification. Accurate NIR in-line particle size analysis of moving granules is challenging, because the scattering and absorptive properties of the granules vary. Nieuwmeyer et al. [7] determined the particle size and the drying end-point of granules using NIR.

10.4.1 Case Study [8]

A Partial Least Square model based on Near Infrared spectra and loss on drying measurements was built by Antonio Peinado et al. [8] to determine in-line, the drying end-point of a fluidized drying process. The product contained active pharmaceutical ingredient (hydrochloride salt) comprising 60−70% of the granule, and contained microcrystalline cellulose, sodium starch glycolate, and povidone. An ABB − Fourier Transform Process Analyzer (ABB-FTPA2000- 260) Near Infrared spectrometer, with thermoelectrically cooled, Indium gallium arsenide (InGaAs) detectors and equipped with a diffuse reflectance probe was used to record in-line NIR data throughout the drying process. Spectra were recorded in-line over the range 1178−2075 nm. Each spectrum was the average of 32 scans. A reference spectrum was obtained at the beginning of each batch. Changes in physical and chemical attributes during drying were monitored using NIR as illustrated in Fig.10.1. The plot shows all the spectra collected during a drying operation. The colorimetric legend is related to the drying time indicating blue to red for the beginning to end of the drying process respectively. The spectral evolution is characterized by the strong absorption around 1940 nm, which corresponds to the combination between the fundamental stretching and deformation vibration of the O−H bond. The researchers concluded that during parallel testing the NIR predictions were closer to

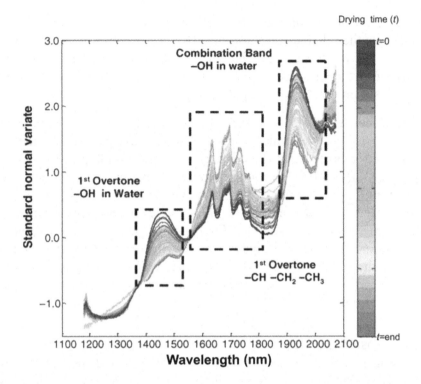

Figure 10.1 NIR spectral evolution for a typical batch during the fluid bed drying process. From Ref. [8].

the end-point specification limit (i.e., not excessively over dried). Therefore, this would have resulted in an average saving in drying time of around 10% if NIR had been used as the primary method for control during parallel testing.

10.5 CHORD LENGTH DETERMINATION (FBRM AND SFV)

Focused Beam Reflectance Measurement (FBRM, (Lasentec by Mettler-Toledo)) and Spatial Filtering Velocimetry (SFV) (Parsum-Malvern), are becoming more recognized in the pharmaceutical industry. Both techniques are designed to directly characterize particle size and track real-time change of particle size and distribution in the process. Although both techniques are laser-based, the measurement principles are different, which lead to differences in the sampling mechanism, measurement range, sample state/conditions, and thus application areas.

10.5.1 Focused Beam Reflectance Measurement (FBRM)

FBRM measures the chord length distribution (CLD) of the particles and chord counts at every minute interval during granulation. The laser beam is rotated at high speed (2−8 m/s). FBRM probe requires that particles flow in front of the probe window while a rotating laser beam is focused on particles. Particles passing near the probe window reflect the laser light and the reflected light is detected. FBRM could detect the three main rate processes (wetting and nucleation, consolidation and growth, and breakage), although the sensitivity of the optical signal was susceptible to fouling of the probe window. A newly developed FBRM C35 utilizes a mechanical scraper to prevent the probe from fouling.

Due to this well-known disadvantage of optical in-line probes, an at-line FBRM application has been developed to enable granule growth studies. Hu et al. [9] studied the chord length distribution (CLD) measured by the FBRM to represent granule particle size distribution (PSD). The study showed that the trends of the chord length measured by the at-line FBRM technique were identical to those measured by a laser diffraction instrument and sieve analysis in spite of different measurement mechanisms. Authors found that the at-line FBRM enables the selection of appropriate process parameters and effectively controls the fluid bed granulation process. Kukec S et al. [10] studied the granule growth kinetics during in situ fluid bed melt granulation process using FBRM. In addition, the usefulness of these techniques during scale-up of melt granulation was evaluated. They concluded that the in-line measurement technique is a viable tool for process monitoring during the transfer of granulation to the larger scale or other manufacturing site/equipment. Alshihabi et al. [11] studied the influence of the position of the probe placed horizontally and at an 45° angle. They concluded that horizontal installation resulted in noisy, high scattering and non-identified curves in the FBRM progression profile due to spray solution sticking and occluding the window, while placing the probe with an inclination angle of 45° allowed air flux to sweep away any particles that adhered to the window and kept it dry during the granulation process and showed smoother and more reliable curves representing the continuous particle growth during granulation. Optimization of the position of the probe is critical to keep the FBRM window clear and to enable the process monitoring.

10.5.2 Spatial Filtering Velocimetry (SFV)

Spatial Filtering Velocimetry (SFV) (Parsum by Malvern) is another method to determine the chord length distribution. FBRM uses the backscattered laser light to calculate particle chord length, while SFV applies the generated shadow (Fig. 10.3). During SFV measurements, particles pass through a laser beam and cast shadows onto a linear array of optical fibers. This results in the generation of a burst signal, which is proportional to particle velocity. As the particles pass through the beam, a secondary pulse is generated by a single optical fiber. In SFV the measured particles are dispersed using pressurized air through the measurement zone inside the probe (Fig. 10.2). The velocity and the chord length size of the particles are measured as they move through the laser beam, hence prohibiting light entrance to the detectors. Knowing the velocity of the moving particle and the time of the pulse, particle chord length is calculated. With the Parsum probe, a laser light obscuration signal from individual particles can be translated into size information for analysis by extended spatial filter as particles pass through an aperture on the probe tip. To minimize fouling of the window, pressurized air is used to disperse particles to minimize fouling. The advantage of the SFV over FBRM is, however, that it can also easily be used as an at-line application without any method

Figure 10.2 Positioning of the Parsum analyzer on the Aeromatic S3 fluid bed. From Ref. [12].

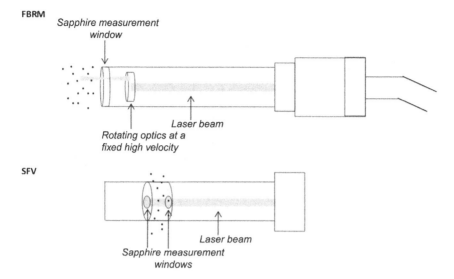

Figure 10.3 FBRM and SFV operating principle. From Ref. [13].

development, and consequently be utilized in granule growth studies in the fluid bed process.

Närvänen [14] evaluated various methods for particle size determination both on-line and at-line methods which included sieve analysis, image analysis, laser light diffraction, SFV, and concluded that most of the methods evaluated can be applied in fluid bed granulation process development. Although there were some challenges in in-line application of SFV, the method itself was proved to be fast, gave reproducible results and concluded that SFV could be useful in process development as an at-line technique.

10.6 THE LIGHTHOUSE PROBE [15]

The Lighthouse Probe (Fig. 10.4) offered by GEA Pharma Systems, uses spectroscopic technology using a Diode Array Detector (DAD). It can be used with a variety of spectroscopic techniques, including NIR and UV/vis, to take reliable in-process measurements of quality-critical product characteristics during processing. The Lighthouse Probe consists of five elements: a probe with optics, a spectrometer, software, movement unit and housing, and clean in-place (CIP) unit. The CIP provides visibility throughout monitoring of the process. The DAD sensor for end-point determination gives more consistency between

Figure 10.4 Lighthouse Probe with a fluid bed processor. Courtesy: GEA Pharma Systems.

batches in granule properties, enabling optimization of subsequent processing steps. It offers a solution for observation window clouding and has the capability of self-cleaning.

10.7 IMAGE ANALYSIS

The characterization using image processing and analyzing techniques includes five steps:

- image acquisition,
- preprocessing,
- segmentation,
- extraction, and
- representation of the characteristic parameters.

One commercially available image analysis technique is QICPIC offered by Sympatec, that uses similar dispersion systems as laser diffraction. The basic concept is the combination of a powerful disperser with dynamic image analysis (DIA). A well-dispersed particle flow is led through the image plane. Due to the dispersion, the particles are separated from each other by the transportation fluid and overlapping particles are widely avoided. So high particle numbers per image frame can be captured (Fig. 10.5). Using the image processing technique, the mass median particle size and the shape factor of the particle could be determined.

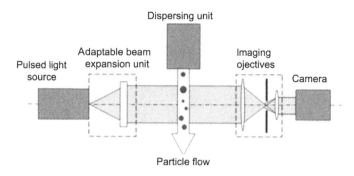

Figure 10.5 Operating principle of QICPIC image analyzer [18].

Watano et al. [16–17] determine the particle size results by the image processing system corresponded well with the sieve analysis results. Study results also revealed that the position of the image analysis probe influences the results due to the particles segregation in fluid bed granulator.

10.7.1 Eyecon [19]

The Eyecon is a particle sizing and characterizing system which can be used both in-line and at- line. Innopharma and Glatt group have teamed up to offer this technology for solid dosage processing. The Eyecon particle characterizer is based on high-speed machine vision. It enables the capture of both size and shape of particles between 50 and 3000 microns. A continuous image sequence of the particles is captured using illumination pulses every microsecond capturing the image of fast moving particles. The illumination is arranged according to the principle of photometric stereo for capturing the 3-D features of the particles in addition to a regular 2-D image. The particle size is calculated from the images using the 2-D and 3-D information, applying novel image analysis methods and direct geometrical measurement. As the approach is based on direct measurement instead of indirect, such as laser diffraction, there is no need for material-based calibration. The Eyecon particle characterizer can be integrated onto most fluid bed granulators with no modification of the equipment, by mounting it outside a pre-existing window (Fig. 10.6).

NIR has typically been a single-point measurement, but last year Innopharma Labs introduced the Multieye, a multipoint NIR

Figure 10.6 Eyecon monitoring fluid bed process. Courtesy: source—Innopharma labs.

spectrometer that can measure up to four points at the same time, which results in a larger sampling area that gives a better representation of the product. From a regulatory perspective, it is easier to demonstrate that multipoint analysis is a representative measurement, according to the manufacturer.

10.8 OTHER METHODS

Acoustic Emission, Raman Spectroscopy, Artificial Neural Network (ANN), Fuzzy logics are some of the approaches that have been used by researchers, and the reader is advised to look up the theory and application in numerous published articles [20–24].

10.9 SUMMARY

In addition to the PAT techniques and scientific knowledge, formal experimental designs can demonstrate enhanced knowledge of product performance over a wide range for material attributes, processing options and processing parameters. It is important to identify parameters that can be changed to influence responses which describe the process. A risk assessment may yield this kind of information. Applications of DoE are screening and optimization studies that ultimately can determine the optimal process region (design space), robustness testing and mechanistic modeling. The control strategy ensures that the operation of the process is maintained within the design space. For optimizing the fluid bed processing, the design space, together with an appropriate control strategy, will reduce end product testing, while increasing process performance and robustness.

With the availability of various options for PAT, process control would help define a robust and optimized process. The selection of various approaches discussed will depend on the criticality of the results needed, and the capability of the technology to deliver the intended results for a given process.

REFERENCES

[1] <http://www.fda.gov/downloads/Drugs/.../Guidances/ucm070305.pdfxx>.

[2] Ikari IL, Martens H, Isaksson T. Determination of particle size in powders by scatter correction in diffuse near-infrared reflectance. Appl Spectrosc 1988;42:722−8.

[3] O'Neil AJ, Jee RD, Moffat AC. The application of multiple linear regression to the measurement of the median particle size of drugs and pharmaceutical excipients by near-infrared spectroscopy. Analyst 1998;123:2297−302.

[4] Frake P, Greenhlgh D, Grierson SM, Hempenstall JM, Rudd DR. Process control and end-point determination of fluid bed granulation by application of near-infrared spectroscopy. Int J Pharm 1997;151:75−80.

[5] Rantanen J, Antikanen O, Mannermaa JP, Yliuusi J. Use of near-infrared reflectance method for measurement of moisture content during granulation. Pharm Dev Technol 2000;5(2):209−17.

[6] Rantanen J, Lehtola S, Rämet P, Mannermaa JP, Ylirussi. On-line monitoring of moisture content in a instrumented fluidized bed granulator with multi-channel NIR moisture sensor. Powder Technol 1998;99:163−70.

[7] Nieuwmeyer FJS, Damen M, Gerich A, Rumini F, Maarschalk KVDV, Vromans H. Granule characterization during fluid bed drying by development of a near infrared method to determine water content and median granule size. Pharm Res 2007;24(10):1854−61. Available from: http://dx.doi.org/10.1007/s11095-007-9305-5.

[8] Peinado A, et al. Development, validation and transfer of a Near Infrared method to determine in-line the end point of a fluidized drying process for commercial production batches of an approved oral solid dose pharmaceutical product. J Pharmaceut Biomed Anal 2011;54:13−20.

[9] Hu X, Cunningham JC, Winstead D. Study growth kinetics in fluidized bed granulation with at-line FBRM. Int J Pharm 2008;347(1−2):54−61 Epub 2007 Jul 3.

[10] Kukec S, Hudovornik G, Dreu R, Vrečer F. Study of granule growth kinetics during in situ fluid bed melt granulation using in-line FBRM and SFT probes. Drug Dev Ind Pharm 2014;40(7):952−9 http://dx.doi.org/10.3109/03639045.2013.791832. Epub 2013 May 13.

[11] Alshihabi F, Vandamme T, Betz G. Focused beam reflectance method as an innovative (PAT) tool to monitor in-line granulation process in fluidized bed. Pharmaceut Dev Technol 2013;18(1):73−84.

[12] Huang J, et al. A PAT approach to enhance process understanding of fluid bed granulation using in-line particle size characterization and multivariate analysis. J Pharm Innov 2010;5:58−68 DOI 10.1007/s12247-010-9079-x.

[13] Burggraeve A, Monteyne T, Vervaet C, Remon JP, Bee TD. Process analytical tools for monitoring, understanding, and control of pharmaceutical fluidized bed granulation: A review. Eur J Pharmaceut Biopharmaceut 2013;83(1):2−15.

[14] Närvänen T. "Particle Size Determination during Fluid Bed Granulation" ACADEMIC DISSERTATION. Finland: Division of Pharmaceutical Technology, Faculty of Pharmacy University of Helsinki; 2009.

[15] <http://www.gea.com/en/products/lighthouseprobe.jsp>.

[16] Watano S, Miyanami K. Image processing for on-line monitoring of granule size distribution and shape in fluidized bed granulation. Powder Technol 1995;83:55−60.

[17] Watano S, Sato Y, Miyanami K. Optimization and validation of an image processing system in fluidized bed granulation. Adv Powder Technol 1997;8:269−77.

[18] <http://www.sympatec.com/EN/ImageAnalysis/ImageAnalysis.html>.

[19] <https://www.innopharmalabs.com/tech/products/eyecontm>.

[20] Parikh DM, Jones D. Batch fluid bed granulation 204−260pp, 2009. In: Parikh DM, editor. Chapter 10, Handbook of Pharmaceutical Granulation Technology. 3rd Edition NY: Informa Health; 2009.

[21] Burggraeve A, Monteyne T, Vervaet C, Remon JP, Bee TD. Process analytical tools for monitoring, understanding, and control of pharmaceutical fluidized bed granulation: A review. Eur J Pharmaceut Biopharmaceut January 2013;83(1):2−15.

[22] Lipsanen T, Narvanen T, Raikkonen H, Antikainen O, Yliruusi J. Particle size, moisture, and fluidization variations described by indirect in-line physical measurements of fluid bed granulation. AAPS Pharm Sci Technol 2008;9:1070−7.

[23] Walker G, Bell SEJ, Vann M, Jones DS, Andrews G. Fluidized bed characterization using Raman spectroscopy: applications to pharmaceutical processing. Chem Eng Sci 2007;62:3832−8.

[24] Halstensen M, Esbensen KH. Acoustic chemometric monitoring of industrial production processes. In: Bakeev KA, editor. Process Analytical Technology. Second ed. United Kingdom: John Wiley & Sons Ltd.; 2010. p. 281−302.

CHAPTER 11

Process Scale-Up

11.1 INTRODUCTION

Current interest in the industry toward continuous processing with the smaller footprint withstanding, commercial products being developed and scaled up represent major processing challenges for processing professionals. Scale-up in the pharmaceutical industry is unique in that experiments at laboratory and pilot scale are also required to produce products of desired specifications for different stages of clinical trials. The appropriate quality management principles can be helpful in prioritizing pharmaceutical development studies to collect knowledge of process performance. In the fluid bed process granule size is directly proportional to the bed humidity during granulation; hence, control of this humidity during scale-up is essential.

The statistically designed experimental designs, such as factorial and modified factorial designs, can generate mathematical relationships between the independent variables such as process factors and dependent variables such as product properties. However, fluid bed processing is a complex process that cannot simply be scaled up by establishing and using only process variables. Identifying and explaining all critical sources of variability, managing variability by the process, and predicting product quality attributes over the design space are indications that a process is well understood. The scale-up from the laboratory equipment to production size units is dependent on equipment design. The material characteristics, such as particle size distribution, density and batch size, are dynamic and will change during both granulation and drying processes.

Because airflow is one of the components of the drying capacity of a fluid bed system, the ratio of air volume per kg or liter of the product is very critical to achieve scale-up that is linear. The other design

How to Optimize Fluid Bed Processing Technology. DOI: http://dx.doi.org/10.1016/B978-0-12-804727-9.00011-9

feature is the cross-sectional area of the product container, and how it has been designed throughout the various sizes that a manufacturer supplies. The relationship between various sizes of the process containers can be utilized to calculate the scale-up of binder spray rate; if the cross-sectional area is designed linearly, then the spray rate scale-up can be linear.

Performing design of experiments (DoEs) at every scale is not feasible. Hence, understanding the key physical transformations and consideration of equipment-independent, "key response variables" for scale-up/down is critical. Design space grows automatically if extensive process variables versus dimensionless or key response variables are used. Design space in terms of scale-independent parameters may provide regulatory flexibility for tech transfer, instead of reestablishing the design space at each scale. A recent offering of a system called "FlexStream"" claims to offer linear scale-up due to its air distributor design and the placement of nozzles [1].

The added value of the application of process analytical technology (PAT) during process development will increase the required understanding of the interrelationships between process parameters and attributes of the produced granules. Based on this knowledge, reasonable specifications for routine manufacturing can be selected and process windows or design spaces for single-unit operations can be implemented. In contrast to conventional in-process controls, in-line PAT measurements allow monitoring of the entire manufacturing process in real-time instead of collecting single-point data at different time intervals. Hence, PAT can be used to establish the process trajectories of successfully manufactured batches.

Generally, once the effect of process parameters is investigated thoroughly at the laboratory scale equipment and optimized, the next stage is to transfer the process to pilot scale and then to production scale. The majority of the studies reported in literature on scale-up concentrates only on the scaling up of the fluidization process. In the studies by the researchers, scaling relationships were proposed to ensure that small-scale and large-scale systems show the same hydrodynamic behavior [2].

Rambali et al. [3] studied the granulation of corn starch and lactose monohydrate cores (200 μm) using an aqueous hydroxypropylmethyl-cellulose binder solution in three sizes of top-spray fluidized beds having batch sizes of 5, 30, and 120 kg. They showed that the granulation processes can be scaled up successfully by keeping the relative droplet size constant in each scale. Hede et al. [4] defined another scaling parameter called the drying force, which indicates the rate of moisture evaporation from the particles. They performed experiments in three different sizes of top-spray fluidized beds having batch sizes of 0.5, 4, and 24 kg and concluded that for a successful scale-up, the drying force parameter together with the droplet size should be kept constant in each scale. This model was used to test different scale-up principles by comparing the simulation results with experimental temperature and humidity data obtained from inorganic salt coating of placebo cores in three pilot fluidized bed scales. In several studies, more complex models [5–9] were utilized.

11.2 VARIOUS SCALE-UP CONSIDERATIONS AND APPROACHES

As you consider scaling up the fluid bed process, some of the approaches and considerations should be reviewed.

11.2.1 Hydrodynamic Similarity

Particle growth in a fluidized bed is closely related to the particle mixing and flow pattern in the bed. This dictates that the hydrodynamics of the scaled bed should be the same as the small unit, indicating hydrodynamic similarity. In bubbling fluidized beds, bed expansion, solids mixing, particle entrainment, granule growth and attrition are intimately related to the motion of bubbles in the bed. Several rules exist for scaling up a bubbling fluidized bed under the condition of hydrodynamic similarity. Fitzgerald and Crane [10] proposed that the following dimensionless numbers be kept constant when scaling up:

Particle Reynolds number based on gas density $\dfrac{d_p u \rho_G}{\mu}$

Solid particle to gas density ratio $\dfrac{\rho_s}{\rho_G}$

Particle Froude number $\dfrac{u}{(g d_p)^{0.5}}$

Geometric similarity of distributor, bed and particle $\dfrac{L}{d_p}$

Where d_p is the particle diameter, u is the fluidization velocity (superficial gas velocity), μ is the viscosity of fluidizing gas, ρ_G is the density of fluidizing gas, g is gravitational acceleration, L is the fluidized bed height [11].

There are also some cautionary notes relating to the minimum scale for the laboratory scale studies. Slug flow, a phenomenon where single gas bubbles as large as the bed diameter form in regular patterns in the bed, significantly reduces solid mixing. It occurs in tall and narrow beds. Stewart [12] proposed a criterion for the onset of slugging.

$$\frac{u - u_{mf}}{0.35\sqrt{gD_F}} = 0.2$$

To ensure that the bed is operating in bubbling mode without risking slugging, the ratio in the equation above must be kept below 0.2. In addition, both the bed height to bed diameter and particle diameter to bed diameter ratios should be kept low. For the pilot fluidized bed, the diameter should be greater than 0.3 m. The proposed similarity rule requires that two conditions be satisfied. The first condition assures a similarity in bubble coalescence. The second assures the similarities in bubble splitting and in the interstitial flow pattern [13].

11.2.2 Fluidization Velocity

In a fluidized bed, pressure fluctuations are directly related to the passage and eruption of bubbles. The velocity of air flowing at the air distribution plate needs to remain constant to ensure the product is fluidizing in a similar way at different scales. The air velocity should be selected such that it is higher than the minimum fluidization velocity of the largest particle, and below the transport velocity of the smallest particle. The most uniform fluidization might be expected to occur when all the particles are of approximately the same size so that there is no great difference in their terminal falling velocities. If the sizes differ appreciably, elutriation occurs and the smaller particles are continuously removed from the system. If the particles forming the bed are initially of the same size, fines will often be produced because of mechanical attrition or because of breakage due to high thermal stresses. Different bed heights do not produce any significant change on the minimum fluidization velocity. Conversely, the density difference between the materials influences the minimum fluidization velocity. A denser material requires a higher superficial gas velocity to start fluidization. Break-up forces also increase with excess gas velocity; hence, the fluidization velocity is best set near the elutriation point for the smaller particles in the bed, thus ensuring vigorous fluidization at the maximum bubble rate, thereby also minimizing the chances of excessive segregation or de-fluidization of large agglomerates [14].

11.2.3 Bed Moisture Content and Evaporative Capacity

Granule growth in a fluidized bed involves three stages: nucleation, transition, and ball growth. The shift from the transition to the ball growth state is primarily dependent on the bed humidity. A larger unit with its higher airflow provides a higher evaporation rate; it is imperative that the drying capacity in the larger unit be maintained such that the bed temperature is like the smaller unit's bed temperature. This can be accomplished either by increased spray rate, increased air temperature, increased airflow, or a combination of these variables to obtain suitable results. If the inlet air temperature and dew point are held constant, scale-up of spray rate becomes proportional to the increase in air volume. Spray rate must be based on the increase in drying air volume, not the increase in batch size.

11.2.4 Bed Height and Weight (Volume)

There is more mechanical stress to the materials as the product bed height is increased, which may result in friable granule breakage, therefore, an increased number of fines. Production equipment is typically designed such that bed depth is increased as well as bowl diameter. The break-up forces increase with bed depth so that deeper beds are for layered growth or denser agglomerates (aspect ratio 2−3) and shallower ones for agglomerate growth (aspect ratio 0.5−1).

Determination of the quantity of scaled-up batch size is decided by the production equipment capacity in the plant. To determine batch size, the bulk density of the product being processed should be known and can be calculated by knowing the volume capacity of the bowl as follows:

To calculate the batch size (X).
If the bowl size 1100 L.
Working capacity of a fluid bed container is normally (range 50%−100%) 80% = 0.8.
And if the bulk density of the product to be processed is 0.4 g/cc.

Then the batch size X can be calculated as $X = 0.8 \times 1100 \times 0.4 = 352$ kg.

Increase in finished product bulk density is related to increase in bed depth, typically approximately 20%, when scaling from pilot to production quantities.

The Wurster coating application having a longer column in the processor could also facilitate increase in batch size and efficiency. If short partition is used, it will allow the bed to disperse prematurely, allowing some of the coating material sprayed through the upbed and spray dried.

11.2.5 Cross-Sectional Area

To keep the exposure to humidity and temperature constant, similar amount of process gas per mass unit of product and time should be available for smaller and larger units. For example, if the measured airflow in the GPCG-5 (manufactured by Glatt) was 250 m³/h you would start at 2500 m³/h in the GPCG-60, or 6250 m³/h in the GPCG-300 as shown in Table 11.1. See also case study describe later in the chapter.

11.2.6 Droplet Size

To reduce the production time in scaling up, the spray rate needs to be increased and thus the atomization air needs to be increased to maintain the same ratio of mass flow rate of liquid to mass flow rate of air, hence the particle size of the granulated product. These changes may affect the particle size, but usually in granulation scale-up, the particle size increases due to the mass effects. Changing either the inlet temperature or atomizing air pressure, can have profound, synergistic effects on particle size and bulk density of the product. Since droplet size is a function of the air to liquid mass ratio, you cannot go beyond its ability to atomize. If the spray rate needs to be increased, the atomization air should be increased to maintain the same droplet size. If you do not have enough air volume for atomization air, then a higher spray

Table 11.1 Batch Determination Based on Air Distributor Cross Section Area				
FB Unit	Capacity (L)	Bowl Screen Cross Sectional Area (m²)	Scale-Up Factor	Airflow (m³/h)
GPCG 5	22	0.0415	1	250
GPCG 60	220	0.4160	10	2500
GPCG 300	1060	1.0382	25	6250
Source: Glatt Literature.				

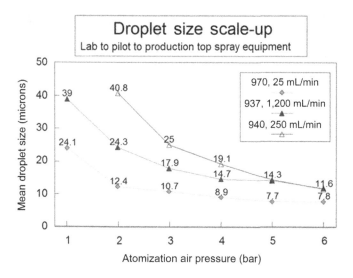

Figure 11.1 Droplet sizes at different atomization air pressure. Courtesy: David Jones, Glatt Group Inc.

rate will not be feasible if the same droplet size is desired. Hence, the droplets should be kept small relative to the size of the substrate particles. For a 5-kg batch, atomization air pressure can approach 40 CFM, while for a 60-kg batch, atomization air pressure can approach 100 CFM to maintain the same droplet size. Fig. 11.1 shows the impact of airflow and resulting droplet sizes. Additional multi-headed nozzles may be required to maintain the same air to liquid ratio and reduce the spray rate at each nozzle port. The recommended operating range for the nozzle is 2.0−3.0 bar.

11.2.7 Key Operating Variables and Response
Table 11.2 summarizes key operating variables and their responses and impact on product characteristics.

11.3 SCALE-UP OF FLUID BED GRANULATION AND DRYING

The processing factors that most affect granule characteristics are rate of binder addition, degree of atomization of the binder liquid, process-air temperature, and height of the spray nozzle from the bed. Since the ratio of bed depth to the air distributor increases with the size of the equipment, the fluidization air velocity is kept constant by increasing the air volume. A major factor, which must be considered during

Table 11.2 Key Operating Variables and Responses in A Fluid Bed		
Key Operating Variables	**Key Response (Output) Variables**	**Product Characteristics**
Airflow rate	Hydrodynamic behavior	• PSD
Distributor plate area		• Granule morphology
		• Granule porosity
Spray rate	Bed moisture and temperature	
Inlet air temperature/RH		
Atomization air pressure	Droplet size distribution	
Number of nozzle	Binder/saturation distribution	
Spray coverage area		

Table 11.3 How to Calculate the Spray Rate for Scale-Up?				
Fluid Bed Unit	**Capacity (L)**	**Spray Rate (g/min)**	**Bowl Air Distributor Cross Sectional Area (ft^2)**	**Batch Size (kg)**
Glatt WSG 5 FB	22	50	0.44	7.0
Glatt WSG 300 FB	1100	?	11.9	350.00
To achieve the same air velocity through the distributor plate in both product containers, the air volume increase *will be related to the ratio of cross sectional areas: 11.9/0.44 = 27. Therefore, 27 × 50 = 1350 mL/min will be the target spray rate in the WSG-300.*				

the scale-up of the fluid bed granulation process, is maintaining the same droplet size of the binder for assuring successful scale-up. It is important to keep the bed moisture level below critical moisture level to prevent the formation of larger agglomerates.

Since higher airflow along with the temperature (drying capacity) in a larger unit provide higher evaporation rate, one must maintain the drying capacity in the larger unit such that the bed temperature is like the smaller unit bed temperature. Nozzle position should always be such that it should cover the powder bed; hence, the location and the number of nozzle ports are important consideration as you scale-up.

Spray rate and atomization air pressure scale-up can be determined by the drying capacity of the equipment which is directly proportional to the cross-sectional area of the air distribution plate rather than by the increase in batch size.

Following example shows how to calculate the spray rate as you scale up, based on the cross-sectional area of smaller and larger fluid bed air distributors (Table 11.3).

11.3.1 Suggested Scaling Rules for Fluid Bed Granulators [11]

Given the understanding of fluidized bed hydrodynamics and granulation rate process, the following guidelines for scaling fluidized bed granulators are suggested:

- Maintain a constant fluidized bed height. Granule density and attrition rate increase with the operating bed height:

$$L_2 = L_1 \qquad (11.1)$$

- If L is kept constant, then batch size scales with the bed cross-sectional area:

$$\frac{M_2}{M_1} = \frac{D_{F2}^2}{D_{F1}^2} \qquad (11.2)$$

Maintain superficial gas velocity constant to keep excess gas velocity, and therefore bubbling and mixing conditions similar:

$$\frac{Q_2}{Q_1} = \frac{u_2}{u_1} = \frac{D_{F2}^2}{D_{F1}^2} \qquad (11.3)$$

- Keep dimensionless spray flux constant on scale-up. This is most easily achieved by increasing the area of bed surface under spray (usually by increasing the number of nozzles). By doing this, the liquid flow rate can be increased in proportion to batch size without changing critical spray zone conditions. Thus, batch times at small and large scale should be similar:

$$\dot{V}_2 = \dot{V}_1 \qquad (11.4)$$

$$\frac{A_{\text{spray},2}}{A_{\text{spray},1}} = \frac{D_{F2}^2}{D_{F1}^2} \qquad (11.5)$$

- Keep viscous Stokes number constant. By adhering to the scaling rules described above, viscous Stokes number (St_v) should automatically be similar at small and large scale leading to similar consolidation and growth behavior.

11.3.2 Additional Rules for Scale-Up of Fluidized Bed Granulators [15]

Granule size distribution and density will not change greatly in scale-up provided the following rules are considered:

- The pilot plant fluid bed unit should be at least 0.3 m in diameters so that bubbling rather than slugging fluidization behavior occurs. Bubbles of similar size and frequency are needed in the pilot and full scale units.
- Where possible, scale-up should maintain constant fluidized bed height. Higher bed will increase density and attrition and will require higher airflow.
- Provided bed height is maintained, gas throughput and solids capacity will scale with bed cross-sectional area. Mass and energy balance limits should be checked on scale-up.
- The type of spray nozzle, droplet size distribution and relative nozzle position should be identical to the pilot unit.

11.4 CASE STUDIES—GRANULATION SCALE-UP

11.4.1 Case Study 1

This example is given as one possible way a granulation scale-up can be done:

Scale-up from Glatt WSG 30 to Glatt GPCG 120 with a batch size scaled from 32 to 107 kg [16]

11.4.1.1 Air Volume Calculations for Scaling Up
1 SCFM = 1.7 m^3/h = 1.7/60/60 = m^3/s

11.4.1.1.1 Glatt WSG 30 Fluid Bed
Air volume in small unit (WSG 30) = Air Velocity in small unit × Screen area × 60
Screen area of WSG unit = 0.194 m^2 (from Glatt literature)
Air volume used for the batch = 500 cfm = 500 × 1.7/60/60 = 0.236 m^3/s
Calculate Air Velocity = Air volume/area = 0.236/0.194 = 1.22 m/s

11.4.1.1.2 Scale-Up to Glatt GPCG 120
Screen area 0.636 m^2 (from Glatt Literature)
Need to find the airflow = x?
To maintain the same air velocity as in the smaller unit = 1.22 m/s
Apply the same formula as above =
Air volume in GPCG = Air velocity × screen area = 1.22 × 0.636 = 0.776 m^3/s
Convert to cubic meter per hour = 0.776 × 60 × 60 = 2794 CMH = 1643 CFM for mixing

Scale up airflow during granulation and drying should be calculated according to the calculations performed above.

11.4.1.2 Spray Rate Calculations
Two ways you can do this:

1. *Air Volume method to calculate spray rate:*
    i. *Air Volume in GPCG 120 divided by Air Volume in WSG 30 × Spray rate in WSG30 = Spray rate in GPCG 120*
    ii. *1765/550 × 300 mL/min (spray rate in WSG 30) = 962 mL/min total*
    iii. *spray rate 2566/800 × 300 mL/min = 962 mL/min*
2. *Cross sectional area method to calculate spray rate:*
    i. *Cross sectional of GPCG/cross sectional area of WSG × spray rate WSG = Spray rate for GPCG = 0.636/0.194 × 300 = 983 mL/min*
    ii. *the spray rate in the larger unit should be between 962 and 983 mL per minute which will require a triple headed nozzle.*
    iii. *The atomization pressure was 2 bar in the smaller and 1.2 mm nozzle port was used.*
    iv. *The larger unit will be using 1.8 mm nozzle port and would require about 3.5 bar air pressure.*
    v. *The volume of air and pressure of air to each nozzle needs to be assured.*
    vi. *If inlet temperature during drying is between 50 and 60°C. It will take approximately 2 h to dry the product, while in the larger unit 2 h of drying might break up granules. Hence, for a larger unit, inlet air temperature needs to be higher to reduce the drying time.*

11.4.2 Case Study 2
Gore et al. [17] conducted scale-up of a batch from 15 to 300 kg and concluded that the appropriate temperature range for the incoming air depends upon the size of the granulator. In a laboratory-size granulator (approximately 15 kg or less), inlet air temperatures of 40−60°C for granulating and of 80−95°C for drying were generally satisfactory. They further concluded, that in a production-size unit (300 kg or larger) in which the powder mass is many times larger, an inlet air temperature in the range of 80−95°C can be used in both the granulating and the drying phases. For a given size of granulator, the inlet air temperature can be varied within the given range to achieve desired granule properties—particularly granule size. On the one hand,

Table 11.4 Case Study 2—Comparison of Process Parameters for Different Size Fluid Beds [17]

Parameters	Glatt WSG-15	Glatt WSG-60	Glatt WSG-300	Glatt WSG 500
Batch size/scale-up ratio	16.0 (1.0)	75.0 (4.8)	375 (23.5)	680 (43.0)
Inlet air temperature granulating (°C)	45–55	60–70	80–90	85–95
Inlet air temperature drying (°C)	85–90	85–90	90	85–95
Binder spray rate (mL/min) (scale-up factor)	240 (1.0)	750–1000 (3–4)	2400–2600 (10–11)	3700–3900 (15.4–16.2)
Moisture content end of granulation	20.4	n/a	n/a	4.0–9.0
Moisture content end of drying	0.8–1.0	0.8	0.8	0.6–1.0
ΔT (°C)	40	n/a		25
Total process time (min)	90	140	170	210
Scale-up ratio	1.0	1.55	2.0	2.3

increasing the temperature should yield smaller and more friable granules because the resulting increased evaporation rate of the binder solvent reduces the contact time, thereby reducing binder penetration and the extent of wetting. If, on the other hand, the inlet air temperature is reduced, the opposite occurs: there is greater penetration and more wetting, resulting in increased granule size and reduced granule friability. Table 11.4 *shows process parameters for different size fluid bed during the study.*

11.4.3 Case Study 3

Matharu et al. [18] *proposed a scale-up equation after scaling up from MP1 to MP6*

A low dose multiple strength product (0.5%–5% w/w active) was scaled from pilot scale using Niro MP-1 Aeromatic Fluid Bed Processor to production scale using Niro MP-6 Aeromatic Fluid Bed Processor. The drug was sprayed using a binder solution (2%) onto a pre-mix of diluent and a disintegrant. The approach to scale-up was based on matching the air velocity between the 2 scales of operations. The impact of droplet size was determined by varying the independent parameters. They investigated the applicability of granulation spray rates to the cross-sectional area of a smaller unit and the larger unit. The underlying phenomenon was based on psychrometry which relates to the total moisture carrying capacity of the incoming air with temperature, volume,

Table 11.5 Case Study 3—Process Parameters Comparison for Two Scales of Fluid Beds [18]			
Description	MP1	MP6	Ratio
Batch size (kg)	8	200	25
Diameter (cm)	17	100	5.89
Air velocity (m/s^2)	1.1	1.1	1
Spray rate (g/min)	51(F1)	1707(F2)	F2/F1 = 33.5
Cross section area (m^2)	0.023 (C1)	0.77 (C2)	C1/C2 = 33.5
Number of nozzles	1	3	3
Inlet air specific volume (m^3/kg)	0.85 ($\delta 1$)	0.955 ($\delta 2$)	$\delta 2/\delta 1$ = 1.13
Inlet air moisture carrying capacity (g)	27 (ΔH1)	16(ΔH2)	ΔH2/ΔH2 = 0.6

velocity and the initial moisture content. The derived equation proposed using the psychrometric, solution flow rate for granulation and a machine dimension of the equipment (cross-sectional area). Table 11.5 *shows the batch sizes and process parameters for the study.*

The relationship between air distributor cross-sectional area and the spray rate is derived as follows:

F1/F2 = C2/C1
Where F1 and F2 are spray rate of smaller and larger units respectively and C1 and C2 is the cross-sectional area of the air distributors respectively.
They introduced in the derived equation, a aerodynamic factor originating from change in machine dimension and thermodynamic factor as follows:
F1/F2 = C2/C1 × Ψ
Where Ψ is inlet air factor defined as $\delta 2/\delta 1 \times \Delta 1 \times \Delta H2/\Delta H1$
F1 is the granulation solution flow rate (g/min) for the lower scale of operation and
F2 is the predicted flow rate for the larger scale of operation. C1 and C2 are the cross sectional areas (m^2) of the air distributors for the two scales of operation. Ψ is the inlet air factor computed from δ (kg/m^3), the specific air volume and ΔH, the humidity ratio (g of moisture/kg of dry air) of the incoming air.
The researchers concluded that the equation could be useful for making scale-up predictions for top spray fluid bed operations. Furthermore, this equation could act as a link between equipment with or without the air handler having dehumidification/humidification provision.

11.4.4 Case Study 4

Product scale-up from 5 to 120 kg in Glatt fluid bed [19]

BATCH SIZE AND EQUIPMENT SELECTION:

$$Smin = [V \times 0.3 \times BD] = [500 \times 0.3 \times 0.4] = 60 \text{ kg} \qquad (11.6)$$

$$Smax = [V \times 0.7 \times BD] = [500 \times 0.7 \times 0.4] = 140 \text{ kg} \qquad (11.7)$$

Where; S is batch size in kilograms;
V is the product bowl working volume in liters;
BD is the bulk density of finished granules in gm/cc;
0.3 = Minimum occupancy of 30% in product bowl;
0.7 = Maximum occupancy of 70% in product bowl;
FLUIDIZATION AIRFLOW SCALE-UP
To maintain the same fluidization velocity, the air volume in a larger unit was increased, based upon the cross-sectional area of the product bowl. In this case, the cross-sectional area of the base of the larger container was 0.64 m² and the smaller was 0.02 m². Thus, correct airflow was calculated as per Eq. (11.8)

$$AF2 = [AF1 \times (A2/A1)] = [80 \times (0.64/0.02)] = 2560 \sim 2600 \text{ CMH} \qquad (11.8)$$

Where; AF1 is Fluidization airflow in the laboratory scale equipment;
AF2 Fluidization airflow in the scaled-up equipment;
A1 is cross-sectional area of the laboratory scale equipment;
A2 is cross-sectional area of the scaled-up equipment;
SPRAY RATE & ATOMIZATION AIR PRESSURE SCALE-UP:
Spray rate scale-up was determined by the drying capacity of the equipment which is directly proportional to cross sectional area of the air distribution plate rather than by the increase in batch size. At a given atomization pressure and airflow volume, change in liquid spray rate directly affects droplet size which in turn impacts particle agglomeration and may cause lumping. Thus, cross-sectional areas of the air distribution plate were used for approximation of scale-up spray rate as per Eq. (11.9).

$$SR2 = [SR1 \times (A2/A1)] = [4 \times (0.64/0.02)] = 128 \sim 130 \text{ gm/min} \qquad (11.9)$$

Where; SR1 is spray rate in the laboratory scale equipment;
SR2 is spray rate in the scaled-up equipment;
A1 is cross-sectional area of the laboratory scale equipment;
A2 is cross-sectional area of the scaled-up equipment;
To maintain the same particle size, the "triple-headed nozzle" in scale-up could use the same spray rate as was used during the development with a similar atomization air pressure. However, this could result in a longer process time. So another approach to maintain a similar droplet size was utilized to achieve mean granule size of 400 µm with maintenance of the mass balance of spray rate and the atomization pressure by increasing the atomization pressure to $2 \times (3) = 6$ bar, the spray rate could be increased to $130 \times (3) = 390 \sim 400$ grams per minute (where 3 indicates number of nozzle heads) keeping the same droplet size and hence obtaining granulation with desired characteristics as required for in-process and finished product quality attributes for larger scale batch. Thus, understanding sources of variability and their impact on downstream processes or processing and finished product quality during pilot scale development stage could provide flexibility for shifting of controls upstream at pivotal scale manufacturing stage and minimize the need for end-product testing and maximize the probability of effectiveness at larger scale.

11.5 FLUID BED COATING SCALE-UP

Fluid bed coating is more challenging than the fluid bed granulation and drying. Coating of powders or particles can be performed using either top-spray where the coating solution is sprayed from the top, bottom-spray (Wurster Technique) where the nozzle is spraying from the bottom or using rotary module with a nozzle spraying from the side or tangentially.

11.5.1 Top-Spray Coating
The conventional top-spray is easiest to scale-up. However, this technique is not suitable for sustained release coating application. Scale-up of spray rate is generally calculated per the increase in fluidization air volume used, not the increase in batch size as discussed previously for scaling up granulation. In fluid bed coating, the most critical factor is to avoid agglomeration due to tackiness of the coating material. This

property determines the maximum spray rate that can be delivered through a single-headed nozzle or coating zone (spray flux). Hence, scale-up requires multiple nozzles increasing the spray flux.

11.5.2 Wurster Coating

Several publications have approached the coating scale-up differently. Mehta [20] suggested scaling air and spray flow rates based on the cross-sectional area for the gas flow in the fluidized bed coaters. Hall [21] concluded that linear scale-up of fluidized bed coating processes is achievable when properly designed air distributor and chambers are utilized and when the air distributor area/fluidizing air volume/spray rate relationships are kept in proportion. In their recent study, Hede et al. [22] used the mathematical model developed by Ronsse et al. [23,24] for batch top-spray fluidized bed coating processes. The model is based on one-dimensional discretization of the fluidized bed into several well-mixed control volumes in which the dynamic heat and mass balances are set up allowing for the simulation of the contents of water vapor, water on core particles and deposited coating mass as well as gas, particle and chamber wall temperature.

Many researchers, especially the ones in industry, followed simpler approaches in modeling and scale-up of the coating processes. Ebey [25] and later Dewettinck et al. [26] developed steady state thermodynamic models for aqueous film coating processes. They utilized these practical models to describe and control the mass and energy balance of the coating process under steady state conditions. Larsen et al. [27] proposed a simpler but dynamic model for process control of aqueous film coating of pharmaceutical substrates.

11.5.2.1 Lab Scale Optimization

Depending on the manufacturer, nozzle configuration for the coating modules varies. A nozzle could be flush with the air distributor (e.g., Precision coater—GEA Pharma Systems) or protrude from the air distributor as is seen in most of the Glatt system and other manufacturer's units. For a given coating application, the choice of what equipment and processing conditions to use must be based on prior experience and availability of equipment in the R & D lab and in production. The fluidization pattern in the Wurster process is very critical. It is advisable that the first experiment in scale-up be performed without a spraying step to visualize the flow pattern of particles through the columns . The Wurster process has number of variables. Some of them are easy to establish, e.g., batch size, spray liquid viscosity, concentration, spray assembly setting, air distributor plate, column height, and dew point, etc. Perform some trials to fix some dependent variables like air volume, atomization air pressure, spray rate, product temperature, etc. Finally apply DoE to optimize the most critical parameters for the process. Before attempting a successful scale-up, key variables and their effect on the output should be identified during lab and pilot scale.

11.5.2.2 Scale-Up

Once the optimized process variables are established, "mass effect" should be considered. It will be easier to compensate for the mass effect by doing minor changes in the predicted parameter in pilot and commercial level when process parameters are optimized. Before the spraying step, load the starting quantity of product in the product container. Observe the fluidization pattern, column height gap, process air volume, and air distributor configuration. Next, add the finished batch size quantity in the product container and repeat the same procedure. During this stage, ideal fluidization conditions should be evaluated at a high process air volume for efficiency and total mass balance. It should be satisfactory across the range of starting and final batch weights.

Product problems such as friability will be magnified as you scale up the process. If the release profile of a modified product is susceptible to the minor processing conditions, scale-up will be a challenge. Therefore, the process developed at the lab scale must be robust. To minimize attrition and reduce the process time, a higher spray rate should be used. With all other variables fixed and after evaluating that particles and coatings do not suffer appreciable attrition, the effect of longer processing time will improve the coating uniformity because of the number of times particles pass through the spray zone. In general, it is better to add a smaller amount of coating to a particle and repeat the process many times than to do the opposite. Air volume from multiple nozzles should be considered when deciding the amount of process air volume that will be needed to evaporate the solution being sprayed.

11.5.2.3 Guidelines for scaling in Wurster Coating

Following are some of the guidelines that can be used to scale-up the coating process using the Wurster module.

- Variables to consider during scale-up (Not all are critical all the time)
  - Inlet air volume, temperature, humidity
  - Nozzle port size & number, spray angle, position
  - Spray rate
  - Concentration of spray solution/suspension
  - Temperature of spray solution/suspension
  - Atomization air volume, pressure, moisture, and temperature
  - Filter area, fabric, porosity, shake cycle & mode
  - Amount of product charged (bed height)
  - Air distributor configuration
  - Wurster column height and gap
  - In the case of rotary fluid bed coating, rotor speed, disc gap, type of plate
- Fluidization pattern during processing is dependent on quantity of product and on air volume. The air volume of scale-up batch is decided based on optimized lab scale batch. From lab to pilot scale the face velocity must be kept the same. To maintain the same velocity one must know the area of the air distributor.
- The volume of fluidizing air can affect agglomeration. During coating, resistance to airflow increases because of outlet filter occlusion, and thus airflow must be adjusted continuously to maintain optimum airflow and fluidization of the beads.
- In the case of controlled release coating, filter bags are replaced by a fine mesh (60 mesh) net or bonnet and shaking of the filter is off.
- Selection of the proper column height in combination with the proper air distributor plate must be monitored to ensure a smooth flow of product in the spray zone.
- In the 18″ Wurster (Glatt units), bed height is up to 600 mm, and fluidization height up to 2 m. Scale-up from 18″ to larger size units with multiple nozzles and columns is almost linear. Hence optimization of the process using 18″ Wurster will provide smoother scale-up to the larger units.
- Since product temperature and dew point are the most critical factors that have an impact on the product movement as well as release profile, during scaling up, these parameters should be kept constant.

- The circulation rate of solids is controlled by the gap between the bottom of the column and the distributor plate. By adjusting this gap, it may be possible to counterbalance the effect of changing particle circulation due to scale-up.
- The base area of Wurster column plays an important role in efficient coating. All process parameters should be proportional to the base area of Wurster column compared with lab model column.
- The following formula can be used to calculate the suitable batch size as you scale-up. This should be a good starting point. The fluidization pattern and the movement of particles through the gap into the spray zone should be taken into consideration as well. In scaling of the process, the height of the product bed increases with increasing batch size. For this reason, a time scale factor consideration is also necessary. Fluid bed batch size determination for coating:

$$Smax = (\pi \times R^2 1 \times H - N \times \pi \times R^2 2 \times H) \times BD$$

$$Smin = \frac{1}{2}(\pi \times R^2 1 \times H - N \times \pi \times R^2 2 \times H) \times BD$$

[where *R1 = Radius of the chamber, R2 = Radius of the partition, N = Number of partitions, H = Length of partition, S = Batch size, BD = Bulk Density*]

- Some typical fluid bed coating capacities vary from 0.5 to 600 kg. For fluid beds up to 450 mm, one Wurster insert (18″) in diameter, is typically used. The diameter of the Wurster insert is nominally one-half the bed diameter. For larger beds, the use of multiple inserts of 225 mm in diameter is common.
- Droplet size scale-up should be based on droplet size data provided by the manufacturer of spray nozzles, but also may be affected by binder properties. Depending on the size of the fluid bed unit, limit the spray rate with low viscosity binders to 1000 mL/min per nozzle port, 750 mL/min with viscous binder solutions.
- Visually achieve same fluidization level inside and outside the partition. Fluidization levels should primarily be adjusted by changing the air distributor pattern. Fluidization control can be done by measuring the product pressure drop. Total airflow and amount of product allowed inside the column (partition gap) can also be varied but may require an adjustment of the spray rate.
- Somewhat higher fluidizing airflows may be needed inside the partition in larger machines as the atomizing airflow is a lower percentage of the total airflow through the partition.

11.6 ROTARY FLUID BED COATING

Of the three fluidized techniques, the rotary or tangential spray system exerts the most mechanical force on any given product. Rotational speed is a key variable and should be kept constant when scaling up the radial velocity used in the lab machine can be calculated by using the formula:

$$Vr = \pi d \times N/60$$

Where V_r = radial velocity in m/s
d = diameter of disc in m
N = number of revolutions per minute the disc is traveling

Knowing the diameter of the disc in the commercial machine and keeping the radial velocity constant, the speed of the rotor (N) can be calculated.

11.6.1 Case Studies Coating
11.6.1.1 Case Study 1

1. *Turton et al. [28] calculated scale-up parameters as follows:*
   Small scale parameters test runs using a 150-mm diameter fluid bed coater indicated that a batch of 2 kg of material could be coated using a liquid spray rate of 10 mL/min for 50 min and a fluidizing gas rate of 40 SCFM.
   It was desired to scale-up this process to a batch size of 200 kg of bed material.
   For the scaled-up process, determination of the bed size, amount of spray solution, airflow rates and the new run time was calculated using Table 11.6 [20].
   Scaled up unit with an 800-mm diameter bed with 3−225 mm diameter columns.
   The airflow and liquid spray rates are scaled on column diameter:
   New airflow required = 40 [3π (225)2/4)/(π (75)2/4) = 40(27)] = 1080 SCFM
   New coating spray rate = 10(27) = 270 mL/min
   The processing time is calculated based on the same coating mass per particle, thus:
   New coating run time = t
   Coating per particle = (270)(t)/(200) = (10)(50)/(2) mL/kg
   Therefore, t = (200) × (50) × (10)/(270)/(2) = 185 min (3.1 h)

Table 11.6 Capacity and Dimensions of Different Size Fluid Bed Units [20]

Bed Diameter (mm)	Column Diameter (mm)	Height of Column (mm)	Number of Column	Approximate Batch Size (kg)
150	75	225	1	0.5–2.0
225	112	300	1	7–10
300	150	375	1	12–20
450	225	600	1	35–55
800	225	750	3	200–275
1150	225	900	7	400–575

Typical airflow rates are on the order of 3−5 m³/s (m² of draft tube).

Table 11.7 Case Study 2—Equipment Set-Up and Process Parameters for Two Scales of Coating [29]

Parameters	GPCG1	CPCG 125 (Pam Glatt)
Wurster column diameter (m)	0.072	0.219
Wurster column height (m)	0.20	0.36
Base plate area (m²)	0.0145	0.1918
Working volume (L)	2.4	84
Batch size (kg)	0.6	21.0
Wurster column base area (m²)	0.0041	0.0377
Gap between air distributor and the column (mm)	15–20	40–45
Inlet air temperature (°C)	26–35	26–35
Product temperature (°C)	26–28	26–28
Spray rate (g/min)	10–20	90–180
Atomization pressure (bar)	1–2	2.5–4.0
Type of nozzle used Schlick	970	940–943

11.6.1.2 Case Study 2

Lab scale batch was manufactured in GPCG 1.1 (6″ Wurster) of pellet core of 200−300 μm which increased up to 500 μm after functional coating (Table 11.7) [29].

11.6.1.3 Case Study 3

Drug loaded pellets, coat with ethyl cellulose for modified release, overcoat with HPMC solution using a 32-inch Wurster to reproduce batches made in 18-inch Wurster. [30]

Following formulation was used (Table 11.8).

Table 11.8 Case Study 3—Formulation Used for Scaling Coating Batch [30]		
Drug solution (For drug layering) Drug 4.4% Binder 1.5% Micronized talc 1.5%	Ethyl cellulose dispersion 18% solid (functional coat)	3. 6.55% HPMC Seal coating solution

Using one batch made in 18-inch as guideline for product temperature, spray rate droplet size and air volume and partition height were projected. The inlet temperature was adjusted to keep the product temperature same in 32-inch Wurster batch. Spray rate per partition was the same as in 18-inch Wurster where, 550 CFM was enough for proper fluidization. Due to equipment differences, 1000 CFM was adequate for 32-inch Wurster. Column gap for 18-inch Wurster was 35 mm, same was chosen for 32-inch Wurster. Two batches were manufactured using these parameters and the results showed mean particle size and the ranges of the two batches were similar. The dissolution data and F2 comparison and product quality was similar.

11.7 SUMMARY

Scaling of granulation and coating requires robust development of the process in the laboratory to identify and optimize the process parameters. Evaluating different lots of active ingredients during the screening process will help the scalability of process. The scale-up requires optimized operating variables identified during the development stages; however, slight variations may have to be made as the process is scaled up. There are various approaches taken by the researchers to determine the parameters for scaling a process. Maintaining the same air velocity, droplet size of the spray solution/suspension and evaporative capacity are critical factors for scaling the fluid bed coating unit operation.

REFERENCES

[1] http://www.gea.com/en/products/flexstream-fluid-bed-processor.jsp.

[2] Glicksman LR, Hyre MR, Farrel PA. Dynamic similarity in fluidization. Int J Multiphase Flow 1994;20:331–86.

[3] Rambali B, Baert L, Massart DL. Scaling up of the fluidized bed granulation process. Int J Pharm 2003;252:197–206.

[4] Hede PD, Bach P, Jensen AD. Top-spray fluid bed coating: scale-up in terms of relative droplet size and drying force. Powder Technol 2008;184:318–32.

[5] Maronga SJ, Wnukowski P. Modelling of the three-domain fluidized-bed particulate coating process. Chem Eng Sci 1997;52:2915−25.

[6] Saleh K, Steinmetz D, Hemati M. Experimental study and modelling of fluidized bed coating and agglomeration. Powder Technol 2003;130:116−23.

[7] Heinrich S, Henneberg M, Peglow M, Drechsler J, Mörl L. Fluidized bed spray granulation: analysis of heat and mass transfers and dynamic particle populations. Braz J Chem Eng 2005;22:181−94.

[8] Degrève J, Baeyens J, Van de Velden M, De Laet S. Spray-agglomeration of NPK fertilizer in a rotating drum granulator. Powder Technol 2006;163:188−95.

[9] Ronsse F, Pieters JG, Dewettinck K. Modelling side-effect spray drying in topspray fluidised bed coating processes. J Food Eng 2008;86:529−41.

[10] Fitzgerald TJ, Crane SD. Cold fluidized bed modelling. Proceedings of International conference of fluidized bed combustion, Vol. III, Technical Sessions; 1985. p. 85−92.

[11] He Y, Liu LX, Litster JD, Kayrak-Talay D. Scale-up considerations in granulation Chapter 25 In: Parikh DM, editor. Handbook of pharmaceutical granulation technology. New York: Informa Healthcare Publishers; 2009. p. 538−66.

[12] Steward PSB, Davidson JF. Powder Technol 1967;1:61−80.

[13] Horio M, Nonika A, Sawa Y, Muchi I. A new similarity rule for fluidized bed scale-up. AICHe 1986;32(9):1466−82.

[14] Nienow AW, Naimer NS, Chiba T. Studies of segregation/mixing in fluidized beds of different size particles. Chem Eng Sci 1987;62:53−66.

[15] Litsrer J, Ennis B. Chapter 9, pp 213 The science and engineering of granulation processes. Boston: Kluwer Academic Publication; 2004.

[16] DMParikh- unpublished data, 1996.

[17] Gore AY, McFarland DW, Batuyios NH. Fluid bed granulation: factors affecting the process in laboratory development and production scale-up pharmaceutical technology, September 1985 and Presented at Pharma Tech Conference 1993.

[18] Matharu et al. (Matharu AS, Patel MR). A new scale-up equation for fluid bed processing. AAPS 2003.

[19] Mukharya A, Chaudhary S, Shah A, Mansuri N, Misra AK. Development and scale-up of SD-FBP formulation technology in line with parametric QbD. Res J Pharm Sci (RAPSR) 2012;1(1).

[20] Mehta AM. Scale-up considerations in the fluid-bed process for controlled release products. Pharm Technol 1988;12:46−52.

[21] Hall HS. Scaling of fluid bed coating, Business Briefing: Pharmatech (2004) 1−5.

[22] Hede PD, Bach P, Jensen AD. Batch top-spray fluid bed coating: scale-up insight using dynamic heat- and mass-transfer modeling. Chem Eng Sci 2009;64:1293−317.

[23] Ronsse F, Pieters JG, Dewettinck K. Combined population balance and thermodynamic modeling of the batch top-spray fluidized bed coating process. Part I—model development and validation. J Food Eng 2007;78:296−307.

[24] Ronsse F, Pieters JG, Dewettinck K. Combined population balance and thermodynamic modeling of the batch top-spray fluidised bed coating process. Part II—model and process analysis. J Food Eng 2007;78:308−22.

[25] Ebey GC. A thermodynamic model for aqueous film-coating. Pharm Technol 1987;11:40−50.

[26] Dewettinck K, De Visscher A, Deroo L, Huyghebaert A. Modelling the steadystate thermodynamic operation point of top-spray fluidized bed processing. J Food Eng 1999;39:131−43.

[27] Larsen CC, Sonnergaard JM, Bertelsen P, Holm P. A new process control strategy for aqueous film coating of pellets in fluidised bed. Eur J Pharm Sci 2003;20:273−83.

[28] Turton, et al. In: Yang WC, editor. Fluidization, solids handling, and processing. Noyes Publication; 1998.

[29] Sonar GS, Rawat SS. Wurster technology: process variables involved and scale-up science. Innovations Pharm Pharm Technol 2015;1(1):100−9.

[30] Bari MM. Wurster coating: scale-up consideration. Bangladeshi Am Pharma Assoc J 2005;39−43.

CHAPTER *12*

Integrated Systems

12.1 MATERIAL HANDLING

The transfer of materials to and from the fluid bed processor is an important consideration. As the process moves from the development through the pilot plant and into commercial operation, material handling becomes a major concern. The loading and unloading of the processing bowl can be accomplished by manual mode, semi-integrated mode or by integrated systems which transfer product to and from the fluid bed depending on the process involved.

12.1.1 Manual or Semi-Integrated Approach

The traditional method for loading the unit is by removing the portable product bowl from the unit, charging the ingredients into the bowl and then placing the bowl back into the unit. This loading is simple and cost effective but labor intensive. Unfortunately, it has the potential of exposing the operators to the dust and contaminating the working area. To avoid the dust and cleaning hazard, a dust collection system should be installed to collect the dust before it spreads. A manual process also depends on the batch size and the operator's physical ability to handle the material. Furthermore, this can be time consuming since the ingredients must be added to the product container, in case of granulation process one ingredient at a time. This approach has the potential of spillage, loss of product affecting yield, besides exposing the operators to dust hazard.

12.1.2 Loading

The loading process can be automated and isolated to avoid worker exposure, minimize dust generation, and reduce loading time. There are two main types of loading systems where the bowl remains in the processor. For the coating process, manual loading and unloading is the most common practice in the industry. In the first option, ingredients to be granulated are added to the bowl by gravity with an overhead integrated bulk container (IBC) without removing the bowl, but the facility should be built with this option in mind, since this requires higher ceiling

How to Optimize Fluid Bed Processing Technology. DOI: http://dx.doi.org/10.1016/B978-0-12-804727-9.00012-0

Figure 12.1 Ingredients from Integrated bulk container (IBC) being transferred to fluid bed. Courtesy: IMA SpA, Italy.

height. The other option is to use the fluid bed itself to vacuum the material from the IBC or from the pre-weighed ingredient containers. The suction of the ingredients from their containers is feasible because an operating fluid bed functions like a vacuum. By connecting the outlet of the IBC to one of the ports of the fluid bed, you can charge the fluid bed. See Fig. 12.1. Once the material has been charged to the fluid bed, the product in-feed valve is closed and the fluidization can continue. This transfer method uses some amount of fluidization air to help the material move through the transfer duct. Loading can be done either vertically from an overhead bin, or from the ground. Less fluidization air is required through the transfer duct when the material is transferred vertically, because gravity is working to help the process. Vertical transfer methods do require greater available height in the process area. Loading by this method has the advantages of limited operator exposure to the product, allowing the product to be fluidized as it enters the processor and reducing the loading time. The disadvantage of this type of system is the cleaning required between different products.

There are a number of reasons why some products are suitable for granulating in the high-shear granulator instead of in the fluid bed processor such as low-dose active pharmaceutical ingredient (API), low

Figure 12.2 Charging product granulated in a high shear mixer. Courtesy: The Glatt Group.

drug content formulations of a hydrophobic drug substance that may have to be incorporated in a larger batch of excipients, product with a very low bulk density, poor flow property of API and/or major excipients, granulated product requiring high bulk density, product with organic solvent binder solution, etc. Most of the time these would be transferred to the fluid bed for drying.

Traditionally in most manufacturing plants a high-shear granulator is either in the same processing room or separately installed in another room. Hence after granulation, the product has to be discharged in a portable fluid bed product container and wheeled in the fluid bed unit for drying. This requires manual handling of the bowl from the discharge point of the high-shear mixer to the fluid bed dryer. For manual handling of wet mass from the mixer, wet granulated product is discharged from the mixer bowl which is placed under the discharge of the mixer and rolled back in the fluid bed processor after collecting the wet mass. Fig. 12.2 shows a portable fluid bed bowl colleting the wet granulated product from the high-shear mixer.

As a more integrated approach, the primary ingredients are transferred from the bulk container into the high-shear mixer, and the discharge from the mixer is transferred to the fluid bed as shown in Fig. 12.3.

Figure 12.3 Shows the ingredients are charged in high shear granulated using the vacuum from the fluid bed. The connection from high shear to fluid bed will be used to discharge the wet granulated product into fluid bed.

Fluid bed coating of pellets or particles requires, in most cases, nozzle spraying from the bottom. When using the Wurster coating bowl for air suspension coating application, the pellets must be carefully loaded so that the product does not enter the column and cover up the nozzles. In most cases, the pellets are loaded manually; even in cases where large batch sizes with three or seven columns. Fig. 12.4 shows the coating container with three columns with pellets properly loaded keeping the nozzle area clear before the product bowl is placed in the processor for coating these pellets.

12.1.3 Unloading

As with loading, the manual method for unloading is by removing the product bowl from the unit. Once the bowl is removed, the operator may scoop the material from the bowl, which is the most time-consuming and impractical method, because of its potential of operator exposure to the product. Alternatively, the product can be vacuum-transferred to a secondary container.

Another option is to have a bowl invertor system installed. The mobile product container of the fluid bed processor is pushed under

Figure 12.4 A three-column Wurster coater with loaded pellets.

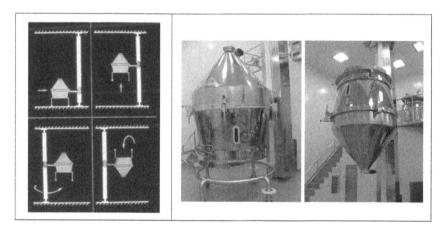

Figure 12.5 Product discharge system inverted product container with a cone mounted on top for discharge or for in-line milling [1].

the cone of the bowl dumper and coupled together by engaging the toggle locks. Subsequently, the container is lifted hydraulically, pivoted around the lifting column, and rotated 180 degrees for discharging (Fig. 12.5). Use of the bowl dumping device or vacuum unloading device still requires that the product bowl be removed from the unit.

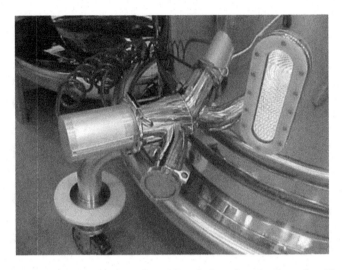

Figure 12.6 Showing the side discharge (capped) and the gill designed air distributor plate (through the site glass). Courtesy: GEA Pharma Systems.

In the case of coated pellets discharge, some manufacturers of the fluid bed-coating module have the capability of tilting the lower distributor plate pneumatically and collect the product in the containers.

There are contained and automated methods for unloading the product while the product bowl is still in the fluid bed processor. The product may either be unloaded out of the bottom of the product container or from the side of the bowl. There are two types of bottom discharge options, gravity or pneumatic. Gravity discharge allows for collection of the product into a container, which is located below the lower plenum probably on the lower floor of the building. If the overall ceiling height limitation prevents the discharge by gravity, the gravity/pneumatic transfer combination can be considered. The gravity discharge poses cleaning problems, since the process air and the product discharge follow the same path; assurance of cleanliness is always of prime concern. The desire to limit the processing area, and development of the overlap gill designed air distributor has prompted the consideration of the side pneumatic discharge as an option. The product bowl is fitted with the discharge gate, as shown in Fig. 12.6.

Most of the granulated product being free-flowing, flows through the side discharge into a container. The remainder of the product is then discharged by manipulation of the airflow through the gill designed air distributor. The discharged product can be pneumatically transported to an overhead bin if the dry milling of the granulation is desired where the mill is installed in-line. The contained system for unloading the product helps to isolate the operator from the product. The isolation feature also prevents the product from being contaminated by being exposed to the working environment. Material handling considerations must be thought of, early in the equipment procurement and facility planning process. Fluid bed processing, whether used as an integral part of a high-shear mixer/fluid bed dryer or as a granulating equipment option, production efficiency, and eventual automation can be enhanced by considering these loading and unloading options to optimize the process.

12.2 INTEGRATED SYSTEMS FOR NONPOTENT COMPOUNDS

For products where the commercial product requirements are high, a manual material transfer will be time-consuming and labor intensive. For dedicated product setup, the fluid bed drying operation integrated with the high-shear granulator equipped with a mill located in the discharge of the high-shear granulator is normally considered. There may be more than one product with a high volume that can be run in an integrated system, but substantial amount of cleaning will be involved. The outlet of the mixer-granulator is directly linked to the inlet of the fluid bed dryer, eliminating long transfer pipes. Thus product contact surfaces are minimized and the yield is increased. Even sticky granules which are difficult to transport by pneumatic conveying can be transferred easily and quickly into the dryer.

If the solvent-based formulations are to be produced in a high-shear granulator, special attention must be given to explosion protection. The integrated systems can incorporate an integrated explosion protection concept. The individual processing units of the mixer and wet mill can be pressurized with a nitrogen blanket while granulating. The dryer will then be designed to be 12-bar pressure shock resistant. The exhaust air leaving the fluid bed processor along with solvent vapors can be controlled in such a fashion that the lower explosion limit of the solvent is not reached. This will depend on how fast the wet mass, laden with solvent, is allowed to enter the fluid bed processor. If the solvent granulation is used, the exhaust air must pass through a catalytic oxidizer to burn off the solvent before exhausting to the environment. Due to the short and direct connection between the mixer and the dryer, the normally required explosion protection valve between the single elements can be omitted. This considerably reduces the cost, set-up times and maintenance.

The mixer-granulator and the fluid bed dryer are installed on a common frame and are connected by an integrated operator platform. Peripheral components such as a wet mill and vacuum conveying system with dry mill are integrated into the plant allowing ergonomic access for inspection, maintenance and cleaning. Due to the optimal ergonomic design, the integrated systems can be set up and operated by one single operator if needed.

Typical integrated systems are shown in the following Figs. 12.7–12.9, where containment is considered for controlling dust and cross-contamination. When these two-unit operations are integrated

Figure 12.7 Integrated system showing high-shear mixer discharge connected through a mill to fluid bed and the discharge from fluid bed connected to overhead hopper. Courtesy: GEA Pharma Systems.

Figure 12.8 Integrated system with a high shear and fluid bed. Courtesy: The Glatt Group.

Figure 12.9 Integrated system with a high shear and fluid bed. Courtesy: L.B. Bohle.

as a single unit, a number of points must be considered. Following is the list of some of the factors the reader may want to consider:

a. Engineering layout and the footprint, ceiling height requirements.
b. How will the high-shear mixer be loaded—by gravity, vacuum, or manually?
c. How will the binder solution be prepared and delivered to the mixer? Is the solvent, water or organic solvent?
d. How will the granulation end point be determined and reproduced?
e. How will the discharge from the high-shear mixer be accomplished?
f. Are the process parameters for granulation and fluid bed drying established and are they reproducible, indicating a robust process?
g. How will the product discharged from the fluid bed dryer be handled? Does it require sizing, blending with the lubricants?
h. Is this system dedicated for a single product or multiple products?
i. How will this system be cleaned?
j. Will the control of a process be done individually for each unit or by an integrated control system?

12.3 INTEGRATED SYSTEMS FOR POTENT COMPOUNDS

In the case of APIs of lower toxicity, the major driving force in design is to prevent the possibility of product contact with the free environment. For APIs of high potency, the major concern in design is to protect the workforce from the hazardous material. This has led to the use of barrier technology, downward laminar flow booths and other containment technology. Innovative design solutions are required to provide practical answers to the problems of containment and cleanliness.

APIs are becoming more and more potent, meanwhile more than 50% of all NCE (New Chemical Entities) are classified as potent (Occupational Exposure Limit (OEL) $<10\,\mu g/m^3$). Furthermore, health and safety authorities all around the world are putting a greater focus on the protection of operators dealing with these substances. In response, suppliers of various hardware components have developed a huge variety of containment solutions, making it difficult to decide which is the best, even for experienced people.

There is a general agreement on a range of containment equipment performances—backed up by occupational health monitoring. There is an increased use of clean-in-place (CIP) systems in facilities, for $<1 \, \mu g/m^3$ compared with $10-100 \, \mu g/m^3$ facilities, accepting trade-off of containment benefits versus increased changeover downtime. The real value of integrated system manifests when potent compounds are to be granulated and dried in a safe manner with minimal exposure to the operators. A facility can be designed to handle multiple APIs with OELs down to $0.01 \, \mu m/m^3$, based on an 8-h time weighted average. The safe manufacture of potent APIs and products containing these APIs requires both "hardware"—facility features, modern equipment and engineering controls—and "software"—programs, practices and procedures—to adequately protect personnel and the environment.

Three main factors dictate how much containment is required and what containment method to consider.

- API quantity, potency and level of toxicity
  The potency of a substance is, in most cases, characterized either by the OEL or by the ADE (Acceptable Daily Exposure). The ADE describes the absolute amount of a specific drug substance that an operator can absorb without any negative health effects. The OEL describes the maximum concentration of a drug substance that can be tolerated in the air of the production room without imparting any negative effect on the health of the operators. For established substances, these values are listed in several textbooks.
- Facility and the process to be executed
  More dedicated facilities are found in those designed for $<1 \, \mu g/m^3$ compared with $10-100 \, \mu g/m^3$. The decision between a multi-purpose versus a dedicated facility is based on commercial or cleanability/GMP reasons, not purely on potency. Companies use different criteria for risk assessment, and there is agreement on the factors considered.
- Work environment, and operational activity of the operators
  Nearly every company has its own classification system for determining the potency of ingredient to be processed, which does not take into account the dilution of the API by excipients. Various companies have different guidelines and procedures for handling potent compounds depending on the facility, product, and the process. While considering the fluid bed processing equipment selection, the equipment supplier, the process team, and safety experts should collectively select how the potent compound will be manufactured and what engineering and operational controls should be implemented.

- To address these factors some of the general guidelines are listed below.
  - Review and document the potential health and safety hazards associated with API including occupational health category and explosivity potential.
  - Determine acceptable exposure levels and measure the potential occupational exposure during the process.
  - Apply containment and controls appropriate for the hazard to reduce risk and verify the ability of the containment to achieve safe and acceptable levels.
  - Determine the potential environmental impact of the API and associated manufacturing processes.
  - Where sealing is not feasible, use barriers and air purging to minimize materials buildup and exposure potential.
  - Expedite decision-making regarding design strategies and containment requirements during technology transfers and during the introduction of new processes.
  - Expedite decision-making regarding design strategies and containment during the design of a new facility or renovation of existing operations.
  - Attain an appropriate degree of containment or infrastructure throughout the facilities for compounds of comparable exposure class or for the same compound to achieve maximum manufacturing flexibility.
  - Implement permanent engineering controls to meet the exposure guidelines.
  - Institute a program of standard operating procedures and training of processing, maintenance and QA/QC individuals.
  - Plan for maintenance requirements.
  - Plan for process upsets and how to recover—a risk assessment must be done. Explore "what if" scenario and document as a procedure.
  - Provide a backup in case of primary containment failure.
  - Provide a controlled flow path through the facilities for chemicals and equipment movement.
  - Separate personnel and material airlocks with decontamination showers in personnel airlocks.
  - Document isolators in potent suites, since paper cannot be decontaminated. Paperless technologies should be evaluated.
  - Continuous monitoring of environmental conditions such as relative humidity; temperature and pressure.

- Once-through air, circulated 20 times per hour, with HEPA supply and return should be engineered.
- Negative pressures in processing areas should be maintained.
- Bag in/bag out HEPA, accessed from the potent process side and/or disposable filters should be considered.
- Contained drain is required.
- Personal protective equipment (PPE) is another costly and involved element of high containment production units. PPE should be required only when isolation by engineering means cannot be made sufficient to protect workers from exposures above compound OELs.

12.3.1 Material Handling for Potent Compounds

Potent compound material handling requires special considerations such as:

Assuring raw material, intermediate, and final product transfer into or out of equipment ports; delivery of components, tools, containers, samples, etc., into or out of an enclosure; connection of critical utilities and/or liquid supply or waste drains; and use of split transfer valves. The transfer system is the key to successful containment. While designing, consideration must be given to cleaning, sampling and maintenance of the system.

A typical contained system with high-shear granulation and fluid bed drying could be set up as follows:

Granulation: integrated line comprising high-shear mixer, integrated wet mill, wet product transfer line from high-shear mixer to fluid bed dryer, fluid bed processor, integrated dry mill and vacuum transfer system for dried granules from fluid bed processor to IBC.
Material handling: IBCs with containment valve, Vibratory feeder and blending prism, disposable high containment interface like Hicoflex which consists of two complementary and self-closing half couplings which seal off dust tight and independently of each other, for discharging API from isolator and charging into high shear, IBC filling station at discharge of fluid bed, post hoist, containment valve on tablet press, IBC wash station.

Fig. 12.10 shows a contained system with through-the-wall design.

Figure 12.10 Fluid bed with through-the-wall design for containment application. Courtesy: GEA Pharma Systems.

12.3.2 Cleaning

During manufacturing solid dosage products contamination/cross contamination is a major concern, especially in high potency situations where minute amounts of compound could produce a therapeutic effect. Stringent cleaning protocols are thus necessary and testing for cleanliness, perhaps down to single-digit ppm or "not detectable" level, is essential. Normally floor sweeping is not permitted and only HEPA vacuum systems should be used. Cleaning the fluid bed and ancillary equipment processing potent compound, CIP system must be used; CIP has several advantages:

- Fluid bed processor with stainless steel cartridges and CIP will eliminate cleaning of the bag filters and exposure to the operator.
- When properly designed and validated, CIP reproduces identical working conditions and provides a better assurance of cleanliness. It is important to remember that manual cleaning cannot be validated.

- An automated system does not require operator presence and can operate on off-shift, saving valuable process time.
- It eliminates the need for disassembly and re-assembly of equipment leading to a shorter plant turnaround times.
- Isolates personnel from toxic or hazardous materials including those used for cleaning.
- It is usually the most economic option as it reduces cost, cleaning time and makes more efficient use of cleaning chemicals. It also reduces effluent and their subsequent treatment.
- Some process material cannot be easily cleaned by just solvent refluxing but a better result could be obtained by the use of detergent and cleaning agents.

12.4 SUMMARY

Technology for integrating high-shear granulator with fluid bed dryer is well established, and has a very high throughput. The granulation process using this method can compensate for fluctuations in raw material specification, and offers the possibility of integration of additional unit operations.

However, there are some limitations such as loss in yield, system requiring a large footprint, higher ceiling height, as well as a long cleaning and changeover for multiple product application. However, CIP that can be validated could help the cleaning effort.

If a new facility is being built to handle potent compounds with an integrated system, a team consisting of experts from the processing, engineering, occupational health, maintenance, quality, safety and industrial hygiene and equipment supplier must meet and come up with a plan of action to address safe, quality potent compound production to optimize the process.

REFERENCE

[1] Parikh DM, Jones DM. Batch fluid bed granulation. In: Parikh DM, editor. Handbook of pharmaceutical granulation technology. 3rd ed. Informa Health; 2009 [chapter 10].

CHAPTER 13

Process Troubleshooting

(Authors Note: The list of problems or challenges mentioned in this chapter are not necessarily the only ones you will encounter, but these are the most common. Your product, your equipment set up, your control system for the unit and the level of expertise of your technical staff would dictate if you will have additional challenges not listed in this list).

13.1 TROUBLESHOOTING THE GRANULATION PROCESS

A. *Poor particle size distribution* (coarse, wet granules mingled with acceptable granules and fines).
   - Probable cause: Spray nozzle performance.
   - Possible solution:
     - Prior to processing of any batch, conduct a functional test of the spray nozzle to assure that it is performing correctly. Poor particle size control and nonuniform distribution of moisture are most commonly the fault of a defective spray nozzle. An effective spray nozzle cleaning/maintenance and testing program is essential. A functional check of the spray nozzle, at the anticipated spray rate and atomizing air pressure/volume must be conducted after a major cleaning. Replacement of the nozzle head (port and air cap assembly) between batches is generally sufficient as a minor clean to assure proper performance. The reason for this is that the "O"-rings and sealing from which the defects originate are in the nozzle body itself. If this component is not disturbed between batches, it is highly unlikely that the nozzle will malfunction during a subsequent batch.
B. *Lumps or larger granules after the process is over*
   - Probable cause: Coalescence of granules—transition into ball growth forming larger granules.

How to Optimize Fluid Bed Processing Technology. DOI: http://dx.doi.org/10.1016/B978-0-12-804727-9.00013-2

- Possible solution:
  - Transition into ball growth is typically seen in the latter stages of the spraying process. Ball growth is indicated by the presence of a considerable number of very large lumps comprising granules, not starting material. Resolution of the problem depends on discerning its onset. The progression of particle size growth is from powder to nuclei to uniform agglomerates. As granule size grows, there is less overall surface area to accumulate the spray liquid. At this stage, the velocity and pattern density also decrease, and there is a tendency for the material in proximity to the spray nozzle to be over-wetted. The excess surface moisture results in coalescence of granules and eventually ball growth. While this is not a common occurrence, it is undesirable and should be mitigated. This problem can be resolved by an increase in fluidization air volume or a slight decrease in spray rate.
  - Because the resulting "balls" comprise porous agglomerates, they may dry reasonable well. However, their size typically leads to a slower moisture loss and consolidation at the base of the product container. As a consequence, they are not seen in the sample port and a final moisture level cannot include their contents. After milling, it is not uncommon for the final moisture content to be higher than that taken at the end of the drying process.

C. *Nonuniform distribution of potent insoluble API (active pharmaceutical ingredient)*
  - Probable cause: Particle size incompatibility—API and excipients.
  - Possible solution:
    - The root cause for the nonuniformity must be identified. A particle size distribution should be conducted and assay can be performed on the various fractions (generally up to 6 sieve sizes) of granulated product. Often the cause is a particle size incompatibility between the API and the granulation excipients, and this will be seen as super-potency in one or more of the sieve fractions. A relatively rigid granule structure at the end of drying and after milling is essential. Examination of the Certificate of Analysis for the API should reveal the particle size distribution, but it says nothing of its shape. Needle-like materials are problematic in that a particle size distribution (using sieve analysis) is a 2-dimensional test

for a 3-dimensional material. A scanning electron microscopy will reveal particle shape and subjectively, the size distribution. If the API is found to be the root cause, either an additional step to bring it into compatibility with the excipients will be needed (such as milling) or the specification to the vendor must be narrowed.

- If the API particle size is very small but the material is cohesive, it is likely that small soft lumps of API remain in the finished granulation. In comparison to high-shear granulation, there is far less mechanical stress in the fluidized bed process. If the API is added as a dry material to other excipients in the product container, it is suggested that it be co-milled with one of the excipients prior to its addition with the remaining materials. The shear of the pre-milling process would be sufficient for de-lumping and would give the mixing process a head start.

- It is common practice at the end of a spray granulation process to shake filter fines into the product container. If this layer is substantial and contains principally very fine material, it should be assayed for potency. If the material is found to be super-potent, the mechanicals for the filter system must be checked. In alternating shaking types of processors, often a gas-tight flap may have lost its ability to seal completely and it must be repaired. A consequence is that there is still air flowing past it during shaking, therefore fines cannot be released from the stiffened filter fabric. This may be externally manifested by a comparatively high filter differential pressure from start to finish in the process. In cartridge filter systems, the effect is similar—material adhering to the filters cannot be released by the compressed air pulse while fluidization air continues through the cartridge. Release is only possible at the end of the batch when fluidization ceases. If this is an issue during process development and scale-up, irrespective of the type of filter shaking, it may be possible to mitigate by trying different types of filter materials.

D. *Low potency of potent API*
   - Probable cause: Poor initial distribution of API; de-mixing of API; preferential retention of API on machine surfaces (expansion chamber, outlet air filter).

- Possible solution:
  - Any residual in the machine tower should be assayed for potency and checked for particle size and distribution to ascertain if it is of primary size or wetted agglomerates. If the material is fine and dry, it is possible that it has de-mixed due to electrostatic charge. This can potentially occur during vacuum charging, or during a product warm-up step prior to spraying if the temperature is high or the step exceeds 1−2 min. In both cases the fluidization air is dry and the environment is fertile for electrostatic charge. If there is considerable residue and it is super-potent, the process can be adjusted such that fluidization forces the granular product into the upper reaches of the expansion chamber and into the outlet air filter to "scour" the residue from these surfaces (during the middle and latter stages of spraying).
  - There is no standardized test for determining either porosity (the size of particle that can be retained) or permeability (quantity of air flow per unit time at a given pressure difference across the fabric). Essentially, one must rely on performance with the product for which it is intended to be used. A production batch (one or more) must be earmarked as "experimental" and processed using the current recipe. If the filter differential pressure is lower, there is some risk that yield will also be less. There is also potential for the API to be lost if it is small in particle size. If this is the case, yet another type of fabric should be tested—the fabric should not dictate process conditions, but must be selected to serve the product and process.
E. *Poor process air temperature control at low process air volume settings*
  - Possible cause: Operation of the machine at too close to the qualified lower limit for temperature and air flow.
  - Possible solution:
    - This is an unfortunate characteristic when the process starts at low air volume and temperature. The air flow sensor accuracy is diminished at low air flows, and the ability of an air handler to control a low temperature at low air flow is an extreme challenge and should be avoided if possible. A higher air volume is recommended even if it results in material being captured in the outlet air filter. If the filter system functions correctly, these fines will be returned regularly to be exposed to the spray liquid, ultimately becoming agglomerates.

Evidence of this is a steady decay in filter differential pressure during spraying.

F. *Bed stalling (in regions of the product container)*
  * Possible cause: High in-process and end-spray moisture content.
  * Possible solution:
    * Experimentation to determine the operating domain (design space) should identify an in-process moisture profile that reaches a failure limit. If this is done and in-process testing includes sampling for moisture, bed stalling would then be seen as a consequence of a breach of this moisture "threshold." A common cause for a sudden shift from success to failure in routine production is calibration of the process air volume sensor. Many fluidized bed spray granulations, particularly those with insoluble raw materials, have spraying conditions in which the air leaving the machine tower is saturated with moisture. The liquid spray rate slightly exceeds the drying capacity of the process air; therefore, the bed builds in moisture. Routine (quarterly or semi-annual) machine calibration always includes the process air volume sensor; of all of the instruments on a fluid bed processor, this is the most difficult to calibrate. Some companies conduct point checks in which the instrument and its transmitter are calibrated while disconnected. Others employ a loop check in which the testing instruments are installed in tandem with the sensor connected in the loop, or a second instrument is used in the duct work to independently confirm the accuracy of the machine indicated value. In either case, if a change is made to the sensor, the user of the processor will not likely see the impact in any of the readings. For example, assume that calibration found the air volume sensor to be indicating a reading that is 5% higher than the actual. When it is corrected, the first batch processed may be found to have in-process and end-spray moisture contents that are higher than usually seen. All of the OIT indicated process parameters are the same as usual, but the batch outcome is different. The problem rests with the air volume sensor (its changed transmitter). If the process operates at saturation, the inlet and product temperature will not change—they represent the condition for each cubic meter or cubic foot of air entering and leaving the batch (at saturation). A sensor found to be off by 5% will mean that less water is being evaporated

per unit time; therefore, the bed is gaining moisture more quickly. If moisture gain is sufficiently rapid, the ball growth or bed stalling threshold may be reached and the batch will be at risk. It is strongly suggested that all calibration data, especially involving changes to ANY instrument be discussed with the equipment users so that the impact of these types of issues can be anticipated and are not "surprises."

G. *Low yield*
- Possible cause:
  - Wrong porosity exhaust filter.
  - Air distributor with coarser screen opening.
  - Filter bag with a tear in it.
  - Filter bag not shaking (or cartridges blow back), duration is too long.
- Probable solution:
  - Make sure the particles size of your API is larger than the micron rating of your filter bag or cartridge.
  - If the product is seen in the lower plenum, loss may be due to the sifting of the material through the air distributor, and may not be visible until the bowl is removed.
  - A tear in the multi-sock filter bag could lose substantial amount of material. This will be evident in the low value of the filter pressure drop. If you have the broken bag detector (which should be a must) installed and calibrated with proper sensitivity downstream from process filters, if there is a tear in the bag, the broken bag detector will provide a signal to shut the machine down.
  - A tear in the bag occurs after several washings when the bags may shrink and tare developed during the installation. So after every washing of the bags, the bag should be inspected before and during installation for any tear.
  - If after checking the filter integrity, and concluding the acceptance, sometimes, if the filter bags are not installed tightly around the filter frame, you may find product, during the fluidization, slip through the straps and end up on the outside of the filter bags. This is not evident until the filter bags are lowered at the end of the process. Assuring that the filter bags are tightly installed around the filter frame should eliminate this problem.
  - Material sticks to the walls of the expansion chamber as a result of static charge the end of the process.

H. *Filter pressure drop reading consistently high*
  - Probable cause:
    - Filters are clogged.
    - Sensors measuring the filter pressure drop is clogged.
  - Possible solution:
    - For measuring the pressure drop (Δp) there are three sensors: one below the product container in the lower plenum, one in expansion chamber, and the third one above the filters. If the filters are clean, and the Δp for the filter still shows a high reading, then the pressure sensor tubing above the filter is clogged up from the product, because this sensor is exposed to any product that may be passing through the filters. When washing the processor, any product in the tubing along with water essentially clogs up the tubing, and shows as high Δp value. Make sure after each processor cleaning, all of the sensors, particularly the filter Δp sensors are clean thoroughly before the filters are placed in the unit.
I. *Excessively coarse granules during granulation*
  - Probable cause:
    - Inlet air temperature too low.
    - High spray rate, due to improper pump calibration.
    - Nozzle position too low.
    - Atomization air is not on and binder does not atomize.
    - Nozzle leakage.
  - Possible solution:
    - Depending on the solvent used, raise the inlet air temperature, so the product temperature is elevated.
    - Make sure the heating source (boiler, etc.) is functioning.
    - Make sure the nozzle position is such that the spray is not localized, and the atomized liquid droplets are covering the bed uniformly.
    - You may have very low atomization air pressure; increase the atomization pressure.
    - The viscosity of the solution may be too high to produce the finer droplet; you can dilute the solution being sprayed.
    - If the nozzle is leaking, it will be evident in erratic fluidization and the product temperature will continue to drop depending on where the temperature probe is located.
    - Lower the spray rate.

- Make sure the solution delivery is calibrated at a regular interval. It is normal practice to calibrate pumps with the binder solution being sprayed prior to spraying.

J. *Excessive fines*
  - Probable cause:
    - Inlet air temperature is too high.
    - Binder spray rate is too low.
    - Insufficient quantity of binder.
    - High fluidization velocity or air flow.
    - Spray nozzle capacity is reached.
    - Binder is not appropriate.
  - Possible solution:
    - Lower inlet air temperature.
    - Optimize the spray rate.
    - Consider using a multiport nozzle if the batch quantity is large enough.
    - Lower the fluidization air volume or use alternate air distributor to reduce the velocity of air.
    - Binder quantity should be increased or stronger binder selected with a high mechanical strength after drying.

K. *Poor fluidization*
  - Probable cause:
    - Too much product in the product container.
    - Incorrect air distributor plate.
    - Processor fan does not have adequate pressure drop.
    - Air distributor not cleaned properly.
    - Exhaust filter porosity is too small.
    - Exhaust filter is blocked.
  - Possible solution:
    - Reconsider the batch size quantity.
    - Make sure the air distributor with a higher velocity (more restrictive air flow) is used.
    - Evaluate the specifications of the exhaust fan capacity and adjust the quantity of product or change the exhaust fan with a higher capacity, and revalidate the system.
    - Make sure cleaning SOP is clearly written and specifically, requires air distributor cleaning and inspection prior to assembling the unit.
    - Exhaust filter may be too restrictive for the particle size of your product and may need to find proper porosity to retain the

product but able to pass the air through. As mentioned above, an "experimental batch" may answer the suitability of the filter.
- Check the filter pressure drop and adjust the filter shake frequency and the duration.

13.2 TROUBLESHOOTING THE DRYING PROCESS

A. *Material transfer from high-shear mixer to fluid bed processor is incomplete in an integrated system*
- Probable cause: There is not enough suction from the fluid bed to pull the material from the discharge port.
- Transfer piping is too long with too many bends or elbows.
- Possible solution:
  - Adjust the fluidization air to create just enough vacuum so the product can transfer in the fluid bed.
  - Depending on the installation of the integrated system, after the product passes from the discharge port of the mixer though the in-line mill, there is a bleed valve that is located in the pipe. Even after adjusting the fluidization air volume, if the product does not move and remains in the pipe, by opening the bleed valve the wet mass in the pipe gets a jolt of extra air, and the material moves.
  - Minimize the distance between the fluid bed and high-shear mixer with in-line observation ports.
B. *Wet mass transferred from high-shear mixer in the fluid bed does not fluidize*
- Possible cause: Wet mass is highly cohesive, or the fluidization air velocity is not sufficient.
- Possible solution:
  - Passing the wet mass through a mill (conical mill) will help break up larger lumps and help in more efficient drying.
  - Increase the air volume avoiding the blinding of the filter (monitor filter pressure drop).
  - Add the wet mass to the drying bowl in small quantities at a time while the fluid bed is running at very slow air volume.
  - Consider integrated system that will connect directly to the fluid bed from the high- shear mixer for controlled feeding.
  - Consider more restrictive air distributor to increase the velocity of the air.
  - A mechanical rake at the bottom the product container may be required to break up large lumps of very cohesive wet

mass. The mechanical action breaks up cohesive lumps, long enough before the fluidization air takes over.

C. *Drying takes too long*
  - Possible cause: Temperature and airflow are too low, incoming air humidity is very high.
  - Possible solution:
    - Increase the velocity of incoming air effectively suspending more product in the air stream.
    - Increase the inlet air temperature (depending on the product).
    - Provide dehumidification system to decrease the dew point.

D. *Product turns into polymorph*
  - Consider drying at a lower temperature.
  - See two examples below where low temperature helps the prevention of transition:
    - Crystal form conversions during the wet granulation process have been reported in the literature. Wong and Mitchell [1] found chlorpromazine hydrochloride converted from form II to form I via an intermediate hydrate phase during wet granulation in an ethanol: water (80.5:22.9 v/v) mixture. The combination of granulating solvent and drying conditions provides a suitable environment for the conversion to alternate crystalline forms.
    - Davis et al. [2] dried glycine granulation at a different rate. The drying rate determined the polymorph content. The faster the granulation was dried, the more rapid the increase in supersaturation with respect to the metastable form, and the greater thermodynamic driving force for the nucleation and crystallization of the metastable form. This was demonstrated by the significantly different polymorph contents from each drying condition. The granulations rapidly dried by fluidized bed drying resulted in more crystallization of a-glycine than the granulations that were tray dried. Additionally, the faster the fluidized bed granulations were dried, the more a-glycine crystallized. By using Near-infrared, the processing conditions, such as the drying rate, can be adjusted to circumvent potential polymorphic transition.

E. *Granules are falling apart after drying*
  - Probable cause:
    - Too high airflow.
    - Improper binder (quality and quantity).
    - High temperature.

- Possible solution:
  - In fluid bed processing, intense mixing and particle to particle collision, particle to vessel wall collision takes place. Breakage and attrition of granules during drying reduces the particle size as the product dries, which generates fines and affects the flow properties of dried product. At the process development stage, the performance of binder may be adequate because of the smaller batch size and the impact of fluidization may not be evident, but as the process is scaled up, the robustness of the bond between the particles should withstand intense fluidization. Formation of fines by attrition or abrasion is in practice an important parameter because it can affect flow property of the granule mass. This breakage depends upon the strength of the formed granule, which in turn reflects the amount and kind of binder is used during granulation. Normally reducing the fluidization velocity at the drying stage will help reduce the generation of fines. Granules can exhibit an intrinsic breakage propensity during drying depending on the water content, type of binder, and extent of stress exposed to the granules.

F. *Final moisture inconsistency*
  - Possible cause:
    - Inadequate process development drying curve.
    - Improper fluidization.
    - Temperature probe out of calibration.
  - Possible solution:
    - Moisture measurement of the product during drying stage is critical. Many older units do not have product temperature probe and rely on the exhaust air temperature probe. Even if the product temperature probe is available, it and the exhaust probe calibration must be checked periodically (at six or twelve-month frequency).
    - Improper fluidization may be due to the air distributor that may be blocked at various locations, which might happen if the cleaning of the air distributor is not properly performed and the previous product residue remains before the new batch is started. Proper cleaning SOP and inspection are required to avoid this situation.
    - During the development of the process, it is a good practice to establish the drying curve by taking samples at various

stages during drying and correlating the change in moisture with the exhaust air temperature, which will help as you move the process in production, and provide a way to determine the end of the drying step.
- Implementation of inline moisture sensors such as NIR probes might help in this situation.

13.3 TROUBLESHOOTING THE COATING PROCESS

A. *Nozzle plugging*
- Probable cause:
    - Suspended solids in the spray line particularly, if suspension is being sprayed.
    - Improperly hydrated polymers with "fish eye."
    - If the methacrylic polymer dispersion is being sprayed, fluidization temperature may be too high.
- Possible solution:
    - This problem is most common while layering (spraying) a drug suspension with a particle size much larger than the nozzle port. It is advisable to assure that the insoluble ingredients in the suspension be micronized prior to making a suspension.
    - Prepared suspension can be milled through the homogenizer if the polymer is not affected by milling action. Alternatively, the insoluble ingredients, such as talc, can be milled separately before adding to the rest of the coating solution.
    - Reduce the distance between the nozzle and the pump station to minimize the possibility of particles settling in line.
    - After any process interruption, the suspension lines should be drained to avoid any settled particles entering the nozzle.
    - Make sure that the nozzle port is large enough compared to the suspended particles.
B. *During Wurster coating the tubing of one of the nozzles experiences back pressure*
- Possible cause: nozzle is plugged.
- Possible solution:
    - The spraying must be interrupted and nozzle should be removed. In a multi-nozzle Wurster set up, the plugged nozzle is easy to identify by the back pressure building up in the suspension line (if the individual nozzle is supplied with separate pump). The nozzle spraying solution may also get plugged

because the substrate material is in the nozzle port and has plugged the nozzle. If the product has to be discharged from the bowl due to nozzle clogging (this is the situation in most older units), and if the machine is equipped with atomizing air volume sensors, just prior to pausing the process, make a note of this volume (total or for individual nozzles). Discharge the batch and repair the defective nozzle. When finished, conduct a test of the spray nozzles (test mode, if applicable) in the empty machine at the desired atomizing air pressure. If the volume equals the value recorded just before the pause, the nozzle annuluses are clear—reload the batch and continue to completion.

- In some of the newer machines, the nozzle access is from the side of the processing bowl and not from the bottom (that is why you have to empty the bowl every time there is a nozzle plug). Because of the easy accessibility of nozzles from the side, the batch does not have to be emptied and the nozzle can be quickly removed; blockage is removed, nozzle cleaned and re-inserted in the nozzle port after the spray test. All these activities can take place in a very short period of time, while the process is still running.

C. *Nozzle tubing is wiggling drastically while spraying*
   - Possible cause: compressed air leaking in the liquid line.
   - Possible solution:
       - If the pump tubing is jiggling erratically, this is a sign that compressed air is leaking into the liquid line. This will cause significant distortion in the spray pattern in particle coating or layering process. If the machine is fitted with a mass flow meter, installed between the pump head and the nozzle wand, the display will be jumpy—a good nozzle should yield variability of only $\pm 2-3$ g/min. about the set point. Erratic pulses exceeding this value are an ominous sign. If seen early in the batch, the process should be interrupted and the nozzle defect corrected. This can be prevented by conducting the functionality tests of each nozzle being used prior to using it for the process.
       - Air bubbles in solution delivery line are sometimes visible for transparent liquid lines that, besides giving erratic jiggling motion, may provide a false reading of how much solution quantity is sprayed in. Tightening the tubing connections

may eliminate the air bubbles and making sure that the liquid feed tube is submerged in the solution tank completely. Sometimes, foamy liquid will cause similar problems; hence, foam formed during solution preparation must be de-foamed by allowing gentle mixing to eliminate the air from the solution prior to commencing spraying.

13.4 SUMMARY

Process troubleshooting relies on collected data. The type of data and the collection interval vary by vendor and user, but the preference is to take the process data reading as much as possible and as frequently as possible. For a fluid bed processor, this would minimally include all temperatures (dew point, inlet, product and exhaust), process air volume, atomizing air pressure and volume and spray rate. Dependent variables should also be collected—product and outlet filter differential pressure, liquid line pressure. In systems with complex air handling units, total air volume, pre-heater temperature, ambient air dew point and dehumidification dew point may be added to the list. This recorded information, is valuable in finding out root causes of the problems. A control system capable of printing out all of the process parameters can identify any problems by studying these printouts. A properly trained staff will eliminate a number of these process-related problems. At the same time, the development of a process must employ a design of experiments to identify the critical process parameters. At the development stage these parameters should be properly challenged to avoid surprises during commercial operation. To be able to address these process challenges, knowing the equipment and its functionality along with training and experience are the prerequisites.

REFERENCES

[1] Wong MWY, Mitchell AG. Physicochemical characterization of a phase change produced during the wet granulation of chlorpromazine hydrochloride and its effects on tableting. Int J Pharm 1992;88:261–73.

[2] Davis TD, Peck GE, Stowell JG, Morris KR, Byrn SR. Modeling and monitoring of polymorphic transformations during the drying phase of wet granulation. Pharm Res 2004;21(5).

CHAPTER 14

Fluid Bed Safety

14.1 INTRODUCTION

From the R&D data, it is possible to get some basic understanding of the process and the ingredients to be processed. As the process is scaled up, it is important to characterize the way the products are handled in the fluid bed unit operation, and the process must be documented and understood to allow the Environment Health and Safety professional to determine the potential exposure pathway from the equipment. In addition to the task performed, it is important to characterize each of the contributing products. Knowledge of the products involved, the speed at which the task is performed, as well as the operating conditions such as temperature, pressure, airflow, Kst (explosion severity) value of the product(s) to be processed, can lead to a clear understanding of the contributing factors to potential operator exposure as you scale up.

The fluid bed process handles a large amount of air. This air, in the presence of fine product dust, poses the potential for an explosion. Dry powders can ignite with sparks created by electrostatic discharges that naturally occur within moving air and powder masses. For an explosion to occur three conditions must exist: an ignition source, a fuel, and oxygen. With an explosion, oxygen reacts with the fuel, releasing heat and gases. If a dust explosion occurs in free space, a fireball of considerable extent arises. If the dust explosion occurs in a closed container, then there is a sudden pressure rise that is mainly decided by the following factors: type of dust, size of the dust, dust/oxygen ratio, turbulence, pre-compression, temperature, shape of the container, and ignition source.

This hazard can be enhanced when using flammable solvents. If sufficient ignition energy (static charge) is introduced, an explosion within the processor can take place. In order to contain these dust or flammable solvent induced explosions, fluid bed processors are normally constructed to withstand overpressure of 2.0 bar, meaning

How to Optimize Fluid Bed Processing Technology. DOI: http://dx.doi.org/10.1016/B978-0-12-804727-9.00014-4

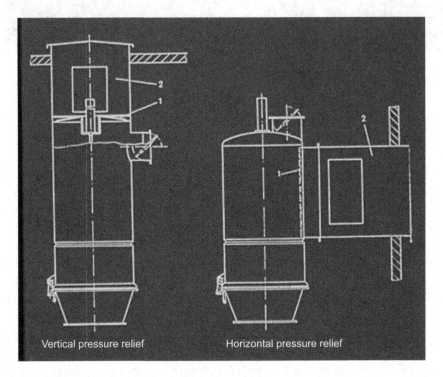

Vertical pressure relief   Horizontal pressure relief

Figure 14.1 Showing vertical and horizontal relief for overpressure.

that the equipment must withstand a two-bar pressure differential across the unit for a short period of time. Two-bar fluid bed units are provided with explosion relief flaps, to release the pressure as soon as it starts to buildup inside the processor. Generally, these relief flaps are located in the vertical or horizontal sections external to the filter housing leading directly to the outside. (Fig. 14.1) These flaps are designed to vent the pressure buildup as low as 0.06 bar.

The 2 bar vented design panels are gasketed and sealed so normal fluid bed operation is not affected. It was accepted practice to have a production unit with 2 bar pressure shock integrity; however, the cleaning of the gasket area around the flaps is always difficult. To avoid having the product exposed to the outside during such an event, a suppression system is used to contain the possible overpressure front from leaving the unit (Fig. 14.2). The suppression system consists of low-pressure sensors located within the processor. These sensors are designed to trigger a series of fire extinguishers (containing ammonium phosphate), as soon as a preset level (generally 0.1 bar) of pressure is set within the processor.

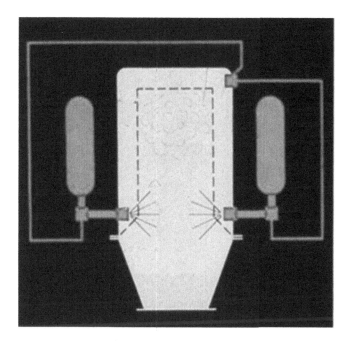

Figure 14.2 Explosion suppression system.

These pressure sensors trigger a series of fire extinguishers, using normally solid quenching materials like ammonium phosphate, as soon as a preset level of pressure is reached inside the machine. With the use of such a system, the external relief flaps must be blocked off to allow for the proper functioning of the pressure sensors.

Units with overpressure capability of 10–12 bar are required if the installation location does not have the possibility of a roof venting or outside venting location, or if the process involves a potent compound and therefore cannot be vented outside in case of explosion. These compounds require total containment of the system (those dealing with highly active pharmaceutical and/or chemical products that are dangerous if not contained).

Most of the pharmaceutical dust explosions studied [1] show the overpressure reaching 9 bar with a Kst value (explosion severity) is at a constant of explosion speed of 200. An explosion in a 10 or 12 bar unit is contained within the unit. A 10–12 bar designed unit does not require any explosion relief panels or gaskets. This eliminates the concerns about cleaning of the gaskets and flaps.

Figure 14.3 Ventex SEI valve (deflagration valve).

Special valves called deflagration valves (Fig. 14.3) are installed in the exhaust duct to contain the explosion. A deflagration valve, such as the Ventex-ESI valve, requires less maintenance than the active valve previously used in the industry. An explosion force (pressure wave) moving ahead of the flame front hurls the poppet forward to the valve seat providing an airtight seal. Once seated, the poppet is locked in by a mechanical shut-off device that retains the seal until manually reset. The three basic versions of the standard mechanical Ventex valve are available with a set pressure of 1.5 psi and a maximum pressure of 150 psi. The Ventex-ESI valve closes by the explosion pressure wave, without external power for horizontal or vertical operation. Fig. 14.4 shows how the Ventex valve closes. The pressure wave of an explosion pushes the closing device against a seal. When closed, the valve is locked and effectively prevents the spread of flames and pressure waves. The actual position of the valve is shown by a position indicator and can be transferred to a control unit via a switch.

For nonsolvent based formulations in fluid bed processors, isolation on the inlet air side is not necessary, because the fluidizing screen of the dryer acts as a mechanical barrier to the propagation of deflagration. However, isolation on the outlet air side is required. For solvent-based formulations, isolation on both inlet and outlet sides is necessary.

In 1994 European Parliament issued an ATEX (*Appareils destinés à être utilisés en ATmosphères EXplosibles*) directive [2] on the approximation of the laws of the Member States concerning equipment and

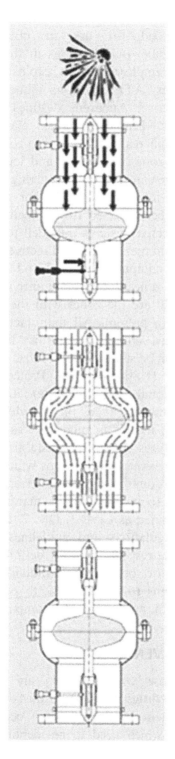

Figure 14.4 Explosion protection valve in action.

protective systems intended for use in potentially explosive atmospheres. As of July 2006, organizations in the EU must follow the directives to protect employees from explosion risks in areas with explosive atmospheres. ATEX gets its name from the French title of the 94/9/EC directive: Appareils destinés à être utilisés en ATmosphères EXplosibles. Employers must classify areas where hazardous explosive atmospheres may occur into zones. The classification given to a particular zone and its size and location, depends on the likelihood of an explosive atmosphere occurring and its persistence if it does. Areas classified into zones (0, 1, 2 for gas-vapor-mist and 20, 21, 22 for dust) must be protected from effective sources of ignition. Equipment and protective systems intended to be used in zoned areas must meet the requirements of the directive. Zones 0 and 20 require category 1 marked equipment, zones 1 and 21 require category 2 marked equipment, and zones 2 and 22 require category 3 marked equipment. Zones 0 and 20 are the zones with the highest risk of an explosive atmosphere being present. All manufacturers of fluid bed processors in Europe must comply with this directive. In North America, the classification system that is most widely utilized is defined by the NFPA (National Fire Protection Association) Publication 70, NEC (National Electrical Code) , and CEC (Canadian Electrical Code). They define the types of hazardous substances that are, or may be, present in the air in quantities sufficient to produce explosive or ignitable mixtures. The NFPA establishes area classifications based on Classes, Divisions, and Groups which are factors combined to define the hazardous conditions of a specific area. When solvents and powders are to be used in the manufacturing area the area classification is designated as Class 1, Div. 2. For US regulations please refer to NFPA regulations and guidelines [3−8]. The local inspection authority has the responsibility for defining Class, Division, and Group classifications for specific areas. Individual countries have developed their own systems to manage risk to personnel, property, production, the environment, and ultimately, company reputation.

14.2 SAFETY FROM SOLVENTS

Special concerns may arise when workers are exposed to toxic solvent vapors and potent drugs as airborne dusts. Worker exposures to solvent vapors and potent compounds may occur during various manufacturing operations, which need to be identified, evaluated and controlled to ensure that workers are protected. Engineering controls are the preferred means of controlling these exposures, due to their inherent effectiveness and reliability. Enclosed process equipment and material handling systems prevent worker exposures, while local exhaust ventilation and personal protection equipment (PPE) supplement these measures. Increased facility and process containment is needed for controlling highly toxic solvents (e.g., benzene, chlorinated hydrocarbons, ketones) and potent compounds. Positive-pressure respirators (e.g., powered-air purifying and supplied-air) and PPE are needed when highly toxic solvents and potent compounds are handled and processed. Special concerns are posed by operations where high levels of solvent vapors (e.g., compounding, granulating and tablet coating) and dusts (e.g., drying, milling and blending) are generated. Locker and shower rooms, decontamination practices and good sanitary practices (e.g., washing and showering) are necessary to prevent or minimize the effects of worker exposures inside and outside the workplace.

Kulling and Simon [9] reported the closed-loop system shown in Fig. 14.5. The inert gas (nitrogen) used for fluidization circulates continuously. An adjustable volume of gas is diverted through the bypassed duct where solvent vapors are condensed and solvent

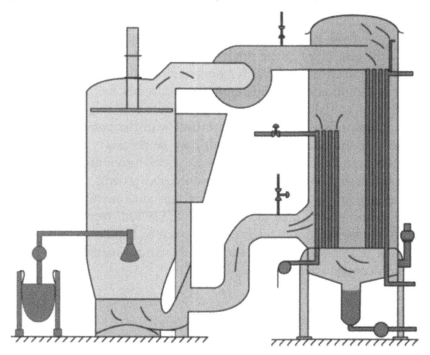

Figure 14.5 Schematic of a closed-loop fluid bed processor with solvent recovery. From [10].

Table 14.1 Comparison of Different Solvent Emission Control Systems [10]

Considerations	Water Scrubbing	Catalytic Burning	Carbon Absorption	Condensation
System	Open cycle	Open cycle	Open cycle	Open cycle with nitrogen
Capital cost	High	Low	Moderate	Low
Energy requirement	High	Low		
Installation	External	External	External	Internal
Space required	Medium	High	Moderate	Small
Flexibility	Medium	Medium	Low	Good
Waste treatment	Required	CO_2/H_2O emission treatment	Required	Concentrated

collected. The circulating gas passes through the heat exchanger to maintain the temperature necessary for evaporation of the solvent from the product bed. During the agglomeration and subsequent drying process, the solvent load in the gas stream does vary. The bypass valve controls the flow of the gas to the heat exchanger and the condenser. By controlling the gas stream in this manner, the drying action is continued until the desired level of drying is reached. Even though the cost of fluid bed processor with the solvent recovery is generally double the cost of a regular single-pass fluid bed processor, such a system offers effective measures for both explosion hazard reduction and an air pollution control. This approach addresses the potential explosion hazard to reduce the amount of oxygen available to the system. Even though this is the best way to deal with the potential explosion hazard, the amount of nitrogen that may be necessary to conduct the complete granulation or drying process is cost prohibitive. In the case of granulation requiring flammable solvents, process air and nozzle atomization air, is replaced by an inert gas such as nitrogen and the system is designed as a closed cycle with the solvent recovery capability [11]. A number of approaches can be taken to handle solvent exhaust from the process. Table 14.1 summarizes various methods for solvent emission control systems.

14.2.1 Safety Precautions

For handling solvents, proper precautions should be taken and risk assessment performed. Following are some of the common sense things to keep in mind while making solvent solutions of binder and processing the granulation step in a fluid bed.

- Minimum quantities only to be used, handled and stored.
- Risk assessments are carried out to identify and minimize the potential for an explosive atmosphere when handling and using flammable liquids.
- Process is carried out in areas classified according to the safety regulations.
- Adequate ventilation is provided where flammables are dispensed, used, or stored.
- All obvious ignition sources are removed from storage and handling areas.
- Electrical items must be safe for use in the zone indicated, or they must be intrinsically safe for use in such areas.
- Nylon lab coats are not used due to potential static problems.
- All flammable liquids are in suitable lidded containers and stored in clearly marked bins or cupboards away from other processes and storage areas.
- Everyone knows the emergency procedure in the event of a significant spillage of flammable liquid—extinguish all flames and heat sources, do not switch electrical appliances on or off, get out and stay out.
- All tanks, pumps, mixers should be grounded.

REFERENCES

[1] Simon EJ. Paper presented at the Third International Powder Technology and Bulk Solids Conference, PowTech Fluid bed processing of bulk solids. Harrogate: Heyden & Sons Publisher; 1975.

[2] ATEX Directive 94/9/EC of the European Parliament and the council, March 23, 1994 and July 2006.

[3] NFPA-68 (current edition): Guide for Venting of Deflagrations.

[4] NFPA-69 (current edition): Standards for Explosion Prevention Systems.

[5] NFPA-70B (current edition): Recommended Practice on Electrical Equipment Maintenance.

[6] NFPA-77 (current edition): Recommended Practice on Static Electricity.

[7] NFPA-499 (current edition): Recommended Practice for Classification of Combustible Dusts.

[8] NFPA-654 (current edition): Standards for the Prevention of Fire and Dust Explosions from Manufacturing Combustible Particulate Solids.

[9] Kulling W, Simon. Fluid bed technology applied to pharmaceuticals. Pharm Technol 1980;4(1).

[10] Parikh DM, Jones DM. Batch fluid bed processing. In: Parikh DM, editor. Handbook of pharmaceutical granulation technology: Chapter 10. Informa Health (Publisher); 2009.

[11] Kulling W. Method and apparatus for removing a vaporized liquid from gas for use in a process based on the fluidized bed principle. US Patent, 4,145,818, March 27, 1979.

Acknowledgments

The author wishes to thank the following companies for providing material and assistance to make this book a success (listed alphabetically).

1. **Bectochem**
   Everest Chamber, A-201, Next to Marol Metro Station, Andheri Kurla Road, Andheri (East), Mumbai 400059, India
   Phone: +91-22-66112100
   E-mail: pharma@bectochemloedige.com
2. **DIOSNA Dierks & Söhne GmbH**
   Am T ie 23 · D-49086 Osnabrück, Germany
   Phone: +49 (0) 541 33 104-0 Fax: +49 (0) 541 33 104-805
   E-mail: info@diosna.com Website: www.diosna.com
3. **Freund Vector**
   675 44th Street, Marion, IA 52302, USA
   Phone: (319) 377-8263 Fax (319) 377-5574
   E-mail: Sales@freund-Vector.com
4. **GEA Pharma Systems**
   9165 Rumsey Road Columbia, Maryland 21045, USA
   Phone: 1 + 410-997-8700
   Website: www.gea.com
5. **GEA Pharma Systems**
   Keerbaan 70, 2160 Wommelgem, Belgium
   Phone: +32 3 350 12 11
6. **Gebrüder Lödige Maschinenbau GmbH**
   Elsener Str. 7, 33102 Paderborn, Germany
   Phone: +49 5251 3090
7. **Glatt Air Techniques, Inc.**
   20 Spear Road, Ramsey, NJ 07446, USA
   Phone: +1 -201 825-8700 Fax: +1 -201 825-0389
   E-mail: info.gat@glatt.com

8. **Glatt GmbH**
   Werner-Glatt-Str. 1, 79589 Binzen, Germany
   Phone: +49-7621 664-0 Fax: +49-7621 64723
   E-mail: info@ glatt.com
9. **IMA S.p.A. - IMA Active Division**
   Via 1 Maggio 14, 40064 Ozzano dell'Emilia (Bologna), Italy
   Phone: +39-051-6514111 Fax: +39-051-6514743
   E-mail: venditeitalia@ima.it
10. **L.B. Bohle, LLC**
    700 Veterans Circle, Suite 100, USA-Warminster, PA 18974
    Phone: +1 (215) 957- 1240 Fax: +1 (215) 957- 1237
    E-mail: info@lbbohle.com
11. **L.B. Bohle Maschinen + Verfahren GmbH**
    Plant 1 Ennigerloh, Industriestrasse 18 59320 Ennigerloh,
    Germany
    Phone: +49 2524 9323-0 Fax: +49 2524 9323-399
    E-mail: info@lbbohle.de
12. **S.B. Panchal and Company**
    #8 Jogani Industrial Estate, 541 Senapati Bapat Marg, Dadar
    (West) Mumbai, 400028, India
    Phone: 91-22-24226882 Fax: 91-22-24226882
    Website: www.sbpanchal.com

INDEX

Note: Page numbers followed by "*f*" and "*t*" refer to figures and tables, respectively.

Printed in the United States
By Bookmasters